MOUNTAINS
AND HIGHLANDS

Tim Harris

Steck-Vaughn Company

First published 2003 by Raintree Steck-Vaughn Publishers,
an imprint of Steck-Vaughn Company.
Copyright © 2003 The Brown Reference Group plc

Library of Congress Cataloging-in-Publication Data

Harris, Tim.
  Mountains and highlands / Tim Harris.
    p. cm. -- (Biomes atlases)
  Contents: Biomes of the world -- Mountains of the world -- Mountain
climates -- Mountain plants -- Mountain animals -- People and mountains.
  ISBN 0-7398-5511-5 (lib. bdg. : hardcover)
  1. Mountain ecology--Juvenile literature. 2. Upland ecology
--Juvenile literature. 3. Mountain ecology--Maps--Juvenile literature.
  4. Upland ecology--Maps--Juvenile literature.   [1. Mountains.
  2. Mountain ecology. 3. Ecology.]   I. Title. II. Series.

QH541.5.M65H37 2002
577.5'3--dc21

                            2002012816

Printed in Singapore. Bound in the United States.
1 2 3 4 5 6 7 8 9 0 LB 07 06 05 04 03 02

The Brown Reference Group plc
Project Editor: Ben Morgan
Deputy Editor: Dr. Rob Houston
Consultant: Dr. Valerie Kapos, Senior Advisor in
    forest ecology and conservation to the
    UNEP World Conservation Monitoring
    Centre
Designer: Reg Cox
Cartographers: Mark Walker and
    Darren Awuah
Picture Researcher: Clare Newman
Indexer: Kay Ollerenshaw
Managing Editor: Bridget Giles
Design Manager: Lynne Ross
Production: Alastair Gourlay

Raintree Steck-Vaughn
Editor: Walter Kossmann
Production Manager: Brian Suderski

Front cover: Mount Cotopaxi, a volcano
in the Andes of Ecuador
Inset: Ibex in the Alps of Europe.

Title page: Mount Roraima, Venezuela.

The acknowledgments on p. 64 form
part of this copyright page.

About this Book

This book's introductory pages describe the biomes of the world and then the mountain biomes. The five main chapters look at aspects of mountains and highlands: climate, plants, animals, people, and future. Between the chapters are detailed maps that focus on major mountain ranges. The map pages are shown in the contents in italics, **like this**.

Throughout the book you'll also find boxed stories or fact files about mountains and highlands. The icons here show what the boxes are about. At the end of the book is a glossary, which explains all the difficult words. After that is a list of books and websites for further research and an index, allowing you to locate subjects anywhere in the book.

Climate

People

Plants

Future

Animals

Facts

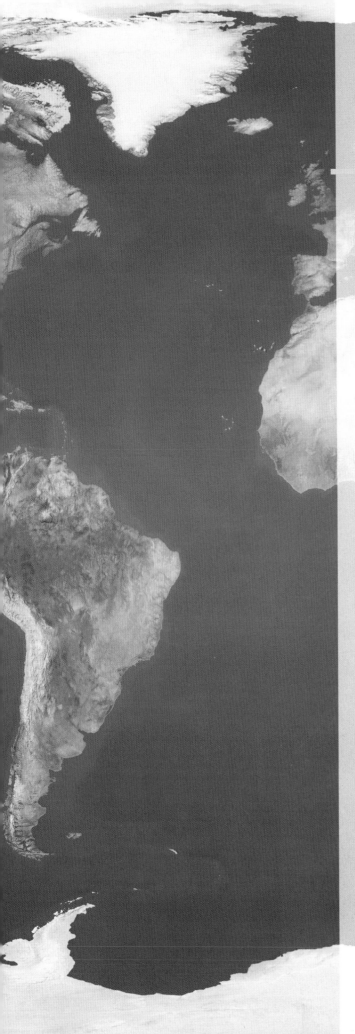

Contents

Biomes of the World   4

Mountains of the World   6

Rocky Mountains   8

Mountain Climates   10

Andes   18

Mountain Plants   20

The Alps   32

Mountain Animals   34

East African Highlands   44

People and Mountains   46

Himalayas   54

The Future   56

Glossary   62

Further Research   63

Index   64

Biomes of the World

Biologists divide the living world into major zones named biomes. Each biome has its own distinctive climate, plants, and animals.

If you were to walk all the way from the north of Canada to the Amazon rain forest, you'd notice the wilderness changing dramatically along the way.

Northern Canada is a freezing and barren place without trees, where only tiny brownish-green plants can survive in the icy ground. But trudge south for long enough and you enter a magical world of conifer forests, where moose, caribou, and wolves live. After several weeks the conifers disappear, and you reach the grass-covered prairies of the central United States. The farther south you go, the drier the land gets and the hotter the sun feels, until you find yourself hiking through a cactus-filled desert. But once you reach southern Mexico, the cacti start to disappear, and strange tropical trees begin to take their place. Here, the muggy air is filled with the calls of exotic birds and the drone of tropical insects. Finally, in Colombia you cross the Andes mountain range—whose chilly peaks remind you a little of your starting point— and descend into the dense, swampy jungles of the Amazon rain forest.

Desert is the driest biome. There are hot deserts and cold ones.

Taiga is made up of conifer trees that can survive freezing winters.

Scientists have a special name for the different regions—such as desert, tropical rain forest, and prairie—that you'd pass through on such a journey. They call them biomes. Everywhere on Earth can be classified as being in one biome or another, and the same biome often appears in lots of

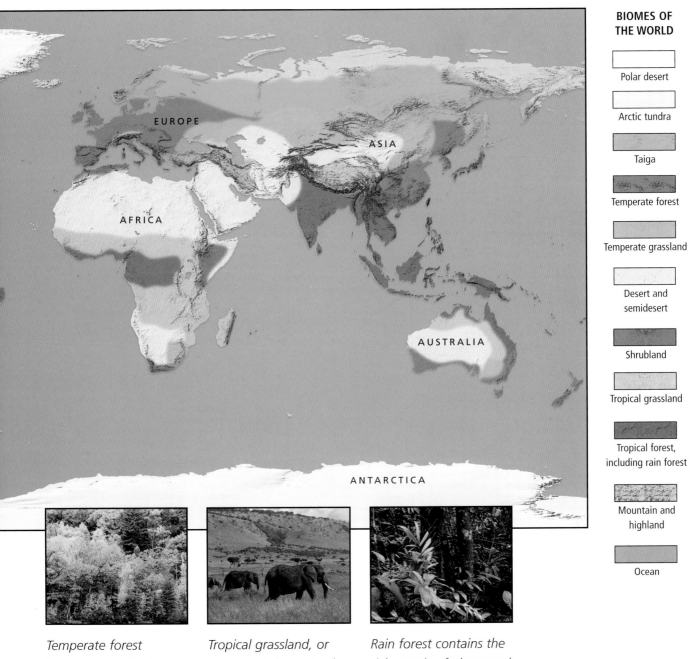

BIOMES OF
THE WORLD

Polar desert

Arctic tundra

Taiga

Temperate forest

Temperate grassland

Desert and
semidesert

Shrubland

Tropical grassland

Tropical forest,
including rain forest

Mountain and
highland

Ocean

EUROPE

ASIA

AFRICA

AUSTRALIA

ANTARCTICA

*Temperate forest
includes trees that lose
their leaves in the fall.*

*Tropical grassland, or
savanna, is home to the
biggest land animals.*

*Rain forest contains the
richest mix of plants and
animals on the planet.*

different places. For instance, there are areas of rain forest as far apart as Brazil, Africa, and Southeast Asia. Although the plants and animals that inhabit these forests are different, they live in similar ways. Likewise, the prairies of North America are part of the grassland biome, which also occurs in China, Australia, and Argentina. Wherever there are grasslands, there are grazing animals that feed on the grass, as well as large carnivores that hunt and kill the grazers.

The map on this page shows how the world's major biomes fit together to make up the biosphere—the zone of life on Earth.

Mountains of the World

On an expedition in one of the world's great mountain ranges, you would see changes in the conditions, plants, and animals as you climbed from the lower slopes to the highest peaks.

Earth's great mountain ranges are so large that they are clearly visible to astronauts orbiting hundreds of miles above us. Some rise miles above sea level—Mount Everest, for instance, is higher than 23 Empire State Buildings. High, mountainous regions can continue without a break over wide areas without rising into sharp peaks, in which case they are called highlands.

There are mountains and highlands on all parts of Earth, so their conditions, their plants, and their animals vary widely. Mount McKinley in Alaska, for instance, rises impressively straight from the arctic tundra at its base, but many of the giant peaks in the tropics have lower slopes swathed in tropical rain forests. Taiga, temperate forest, shrubland, and grassland cover other mountains, and one of Earth's driest deserts, the Atacama, extends into the mighty Andes mountain range in South America.

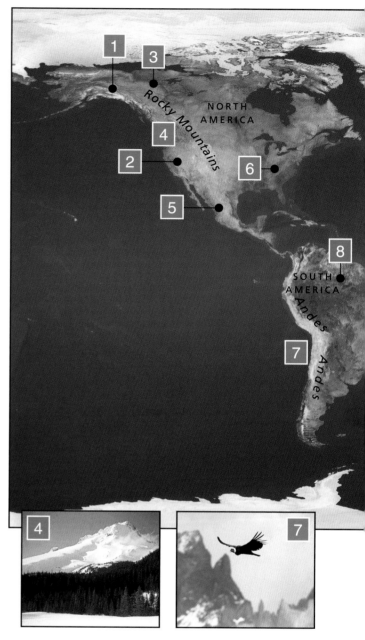

The varied Rocky Mountains stretch from Alaska to Mexico.

The Andes separate desert on one side from forest on the other.

So what do mountains have in common? For one thing, their summits poke up into the upper atmosphere where the air is thin and winds are very strong. Air temperature also decreases as you go higher—everywhere. Climate, plant life, and animal life all change as you climb or descend a mountain.

MOUNTAIN RANGES

1. Alaska Range
2. Sierra Nevada
3. Mackenzie Mountains
4. Rocky Mountains (*see pages 8–9*)
5. Western Sierra Madre
6. Appalachian Mountains
7. Andes (*see pages 18–19*)
8. Mount Roraima
9. Atlas Mountains
10. Alps (*see pages 32–33*)
11. Kjölen Mountains
12. Carpathian Mountains
13. Ruwenzori and Virunga mountains (*see pages 44–45*)
14. Ethiopian Highlands (*see pages 44–45*)
15. Caucasus Mountains
16. Elburz Mountains
17. Zagros Mountains
18. Himalayas (*see pages 54–55*)
19. Tian Shan
20. Kunlun Shan
21. Tibetan Plateau
22. Altai Mountains
23. Mount Kinabalu
24. Maoke Mountains
25. Verkhoyansk Range
26. Japanese Alps
27. Great Dividing Range
28. Southern Alps

The Alps of Europe gave their name to alpine tundra and alpine meadow.

The plants of East Africa's highlands include unusual giant lobelias.

The Himalayas are so high that their peaks are barren wastes cut by glaciers.

At the top of Earth's highest mountains, the Himalayas, you would need to wear a breathing apparatus because the air pressure is so low. No animals or plants live here; conditions are too cold for all but the very simplest life-forms. Lower on the mountainsides are grassy slopes, bright with flowers in spring. Lower still, you would find yourself in forest, with birds, mammals, and other forms of animal life. Yet you would still be thousands of feet above sea level, on the same mountain but in different conditions. Each mountain is not so much a biome as a complex mixture of different biomes.

Rocky Mountains

Pine forests, flower meadows, deserts, glaciers, and even temperate rain forests are just a few of the diverse landscapes that exist in the Rocky Mountains.

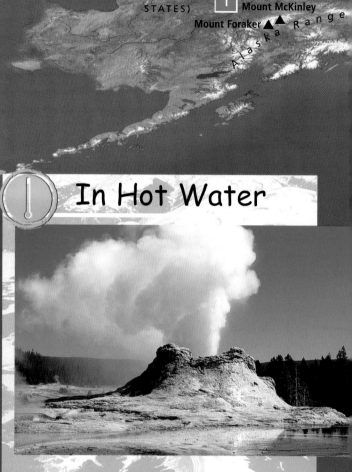

Brooks Range

ALASKA (UNITED STATES) Denali National Park 1 2 Mount McKinley Mount Foraker ▲▲ Alaska Range

In Hot Water

Mount Rainier in Washington State is a dormant volcano surrounded by glaciers and wildflower meadows. It last erupted 2,000 years ago.

Yellowstone is the oldest national park in the world. Established in 1872, this big protected area straddles a colossal volcano. Although dormant, the volcano generates enough heat to boil undergound water and blast it into the air in geysers (such as Castle Geyser, above). The biggest is the Steamboat Geyser, which sometimes rises 300 feet (90 m) and showers viewers with mineral-rich water. The Steamboat is not the most famous geyser; that award goes to Old Faithful. There are more geysers in Yellowstone than anywhere else on Earth.

Yellowstone National Park is also famed for its animals, which include bison strong enough to toss a man into a tree, and grizzly bears, which should be avoided at all costs. People have lived in the Yellowstone area for at least 12,000 years.

NUNAVUT

NORTH AMERICA

SOUTH AMERICA

miles km
400
400

N

Mackenzie Mountains

YUKON TERRITORY

NORTHWEST TERRITORIES

Mount Sanford
Mount Blackburn
Mount Bona
Mount Logan
Mount St. Elias

St. Elias Range

Coast Mountains

Mount Fairweather

3

BRITISH COLUMBIA

R o c k y   M o u n t a i n s

ALBERTA

SASKATCHEWAN

MANITOBA

10 Jasper National Park

11 Banff National Park

C A N A D A

Vancouver Island

Cascade Range

WA

Mount Rainier

MT

ND

Mount St. Helens

Salmon River Mountains

SD

6 OR

Great

ID

WY

12

NE

Yosemite

8

5

Basin

Colorado

UT

COLORADO

13

Coast Ranges

Sierra Nevada

NV

Mount Harvard

KS

4

Plateau

Mount Blanca

CA

9

Mount Whitney

AZ

NM

OK

7 Grand Canyon

TX

1. Mount McKinley
North America's tallest mountain is 20,320 feet (6,194 m) high. Many people have died trying to climb it.

2. Denali National Park
Grizzly bears, wolves, Dall sheep, and moose live in this Alaskan preserve around Mount McKinley.

3. St. Elias Range
This coastal range boasts many of North America's tallest mountains (including Mount Logan, Canada's highest peak) and the world's most extensive ice fields outside the polar ice caps.

4. Sierra Nevada
The picturesque mountains of this California range cast a rain shadow over Nevada, giving the Great Basin a very dry climate.

5. Yosemite National Park
A beautifully scenic valley in the Sierra Nevada range, featuring dramatic cliffs, waterfalls, and sequoia groves.

6. Mount St. Helens
A huge volcano that erupted explosively on May 18, 1980.

7. Grand Canyon
This breathtaking chasm formed over thousands of years as the Colorado River slowly wore its way down into the rising Colorado Plateau.

8. Great Basin
A region of dry valleys and mountain ranges that run from north to south.

9. Mount Whitney
The tallest U.S. mountain outside Alaska, at 14,495 feet (4,418 m). It lies in Sequoia National Park among some of Earth's oldest and tallest trees.

10. Jasper National Park
Visitors to this Canadian reserve can see cougars (mountain lions), black bears, coyotes, moose, and elk.

11. Banff National Park
Rockclimbing, hiking, skiing, and birdwatching are all popular with visitors to this mountain park.

12. Colorado
This state has many of the United States' tallest mountains outside Alaska. It also has some of North America's best ski resorts.

13. White River National Forest
A hikers' paradise in the heart of the Colorado Rockies.

Fact File

▲ The Rockies run all the way from Alaska to New Mexico, forming the backbone of the Western Cordillera, a vast system of mountains and highlands in western North America.

▲ The Western Cordillera began to form about 80 million years ago as the bed of the Pacific collided with North America, making the land crumple.

▲ There are more protected areas in the Rocky Mountains than in any other mountain range.

▲ Some parts of the Rockies receive very little rain, but the coastal ranges in Alaska and British Columbia are extremely wet.

Mountain Climates

Mountains are places of extremes. Some of the wettest, coldest, and windiest places are on mountains, but some mountains are sunny and dry. The weather on a mountain can change drastically in minutes.

Most high mountains, particularly those far from the equator, are cold places. Anywhere on Earth, as you climb higher above sea level, the air temperature falls. For every 1,000 feet (300 m) you climb up a mountain, the temperature gets about 3.4°F (1.9°C) cooler. At the top of Earth's highest mountain, Mount Everest, the temperature is 100°F (55°C) colder than it would be at sea

Fact File

▲ In 1998, Hurricane Mitch dropped up to 75 inches (1,900 mm) of rain on mountains in northern Honduras in just 60 hours—that's twice as much rain as the country's capital city gets in a year.

▲ Mountains experience the highest winds. In 1934, instruments on Mount Washington in New Hampshire recorded a wind of 231 mph (372 km/h).

▲ Mount Rainier, Washington, averages 700 inches (17.8 m) of snowfall a year. That's 58 feet!

level. The summits of all the very highest mountains, even those in the tropics, are permanently covered with snow and ice. Meteorologists (weather experts) use a variety of instruments to measure the weather, but it is difficult to maintain weather stations on high mountains since they are often very remote. As a result, there is still much to learn about mountain climates.

Cold Comfort

Generally, the top of a mountain is colder than halfway up, and halfway up is cooler than the base. However, altitude (height above sea level) is not the only thing that

On mountain peaks, the wind is often so fierce that it blows away clouds of fallen snow. This is the volcano Cotopaxi, which, although situated near the equator in Ecuador, is so high that it is always covered in snow and ice.

controls temperature. If a mountain is in the tropics it is likely to be warmer than a mountain in a temperate or polar region, because the tropics receive stronger sunshine than the poles. For example, average afternoon temperatures at tropical Bogotá, which is near the equator in the Colombian Andes, reach 67°F (19°C); at the South Pole, which is about as high as Bogotá, summer temperatures peak at around –15°F (–26°C). In New York, the average afternoon temperature in June is 80°F (27°C), the same as Kathmandu in the Himalayan foothills. Kathmandu is much higher than New York, but also much closer to the equator.

The direction the mountain slope faces (its aspect) is also important. Slopes facing the sun during the hottest part of the day are much warmer than those in the shade. In the

Downslope Winds

Winds blowing over mountain ranges can produce very unusual weather. As air climbs a mountain, it cools and drops its moisture as rain or snow. By the time it reaches the top, the air is drier and colder. Then, as it descends the far side of the mountain, it becomes squeezed and warms up. This flow of warm, dry air—called a downslope wind—brings clear blue skies and sunny weather. It can also cause unusually hot weather. In June 1995, in the states of Washington and Oregon, a downslope wind from the Cascade Mountains created a heat wave. As it flowed west down the mountains, the air became warmer, ensuring clear skies. Temperatures between the mountains and the Pacific Ocean were much warmer than forecast, reaching 95°F (35°C) at Portland, Oregon—a good 22°F (12°C) above the usual afternoon peak for that time of year.

11

northern hemisphere, southwest-facing slopes are the warmest; in the southern hemisphere, northwest-facing slopes are the warmest. Some steep-sided mountain valleys rarely receive any direct sunshine, and they stay cold most of the time. The amount of sunshine that shines on a mountain slope has a big influence on the plants that grow there, since plants require warmth and light to grow. Plants on sunny slopes grow bigger than those on cool, shady slopes.

At night, temperatures fall but they do not fall evenly. Cold air is heavier than warm air and flows downhill, collecting in hollows or valleys. Overnight, mountain valleys can become colder than mountaintops, the opposite of conditions during the day. This situation is called a temperature

Being enshrouded in cloud nearly all the time is an unusual kind of climate not often experienced outside mountains. In the highlands of southern Venezuela (below), cloud forests grow in the perpetual moisture.

inversion and is why valleys in mountain areas can become frosty or filled with mist in the mornings.

Catching the Rain

Mountains and highland areas are usually wetter than lowlands. There is a simple explanation for this. Since mountains form an obstacle to the winds blowing around big weather systems, the only way the wind can get past is to blow over the top. Air is like a sponge and can hold large quantities of water vapor (a gas formed when water evaporates) that it picks up as it passes over the oceans. Warm air is a better sponge than cool air; it can hold more water vapor. As the air rises it cools, and the water vapor condenses into cloud droplets. Eventually, the water droplets become large enough to fall as raindrops or snow. That is why mountain slopes facing into the prevailing wind (the most common wind direction) tend to be wet, particularly if the wind has blown over an ocean.

Mountain Deserts

The Chilean Altiplano, a highland part of the Atacama Desert in Chile, is probably the driest area on Earth. Although the town of Calama is more than 8,000 feet (2,450 m) above sea level, it never rains there. One explanation for the Altiplano climate is the warm, wet winds that blow west across southern South America. These winds drop almost all their moisture as they climb the Andes mountain range. As the winds descend on the Chilean side of the mountains, the air becomes warmer so it holds onto the little remaining water vapor. The Atacama Desert is said to be in the rain shadow of the southern Andes.

If you look at an atlas showing mountains and prevailing wind direction, you can predict where the wettest places will be. A combination of high mountains near an ocean, with onshore winds, is almost guaranteed to produce a lot of rain over the mountains. Three of the wettest places on Earth are in highland areas of Colombia, Hawaii, and northeast India. From April to October, warm and very wet southwesterly winds blow from the Bay of Bengal over Bangladesh. These winds are known as the southwest monsoon, and since the air is so warm it absorbs millions of tons of water from the ocean. When the air is pushed up the southern slopes of the Himalayas over places such as Cherrapunji, it cools and releases the water in torrents of rain. Something similar happens in western Colombia and on Mount Waialeale, on Kauai, Hawaii, where very warm, moist air that has blown over the Pacific Ocean is forced upward by mountains.

So does it get wetter the higher you climb up a mountain? No—rainfall reaches a maximum at a certain height and then starts to decline, because the air has released most of its water vapor. The height at which this

happens varies from mountain to mountain, but, generally, rainfall begins to decline above 10,000 feet (3,050 m). Mountains that are higher than that are often shrouded in cloud, but rainfall tends to be less.

Rain Shadows

Mountains are not always wet. On the side of a mountain range facing away from the prevailing wind, conditions tend to be drier; the onshore wind has lost most of its moisture on the climb up the mountains, so there is little water to deposit as it descends. Also, as the air descends, it warms, so its ability to hold onto its water vapor increases. It is as if the sponge is getting bigger again. The drier area behind a mountain range is termed the rain shadow. The Sierra Nevada in California, for instance, casts a huge rain shadow, creating semidesert conditions in the mountains and valleys of the Great Basin.

Mountains are also less rainy if the prevailing wind blows across land rather than sea—there are less opportunities for the air to soak up water vapor—or if the wind is cool before it reaches the mountains. This may happen if it has blown over a cold ocean, for example. Since it is cool, it won't be a very effective sponge. The weather station at the South Pole records less than 1 inch (25 mm) of snow a year, despite being about 10,000 feet (3,050 meters) above sea level in the middle of the very high Antarctic ice cap.

In the Teeth of the Gale

In lowland areas and on the low slopes of mountains, winds tend to be slowed by rubbing against the ground. Higher up, they blast into mountains with their full force. In a storm, a mountaintop is a scary place to be. In many cases, people can be blown over by the force of the wind. Mountain valleys can

Seen from a space shuttle, the rain shadow cast by the Himalayas is very obvious. The south side of the mountains (top) is covered with dark forest, but the north side (bottom) looks almost as dry as a desert.

Much more snow falls on some cold mountains than in the cold polar regions. In Yellowstone National Park (right), the snow builds up into huge drifts.

channel and concentrate winds, making them even more powerful. While the side of a mountain facing the wind tends to get lots of rough weather, the lee side (the side facing away from the wind) is relatively sheltered. Trees and shrubs grow taller on the lee side of mountains, animals find more shelter, and people are more likely to live there.

Avalanches

Snowfall is greater in some mountains than in any of Earth's biomes, but it does not snow on all mountains, and even some of the highest mountains have snow only near the summit. For snow to fall, the air must be cold, and there has to be moisture in the air. If the air is too warm, it will rain instead of

Mount Everest

N

Rainy Record Breakers

The windward slopes (above) of the island of Kauai, Hawaii, are drenched with rain. Mount Waialeale, a 5,208-foot (1,588-m) peak on the island, is a strong contender for the rainiest place on Earth. It averages 460 inches (11.7 m) of rain every year, but is eclipsed by Cherrapunji, a village in northeast India. Cherrapunji is more than 4,000 feet (1,300 m) high on the southern slopes of the Himalayas. Average rainfall is 500 inches (12.7 m), with a world record 1,042 inches (26.4 m) in one very wet year. In the rainy season, more rain falls in one month in Cherrapunji than falls in a year in lowland tropical rain forest. There are two reasons why Cherrapunji is so wet: For six months of the year it is drenched by the monsoon rains that sweep up from the Bay of Bengal; and since it is on a mountain, the winds are forced to drop their moisture there.

snowing; if the air is too dry, there will be neither rain nor snow. The line above which snow remains on the ground all year is termed the snow line; this level is higher in the tropics than in temperate latitudes.

Snow can be great fun; thousands of people travel to mountains each year to ski, toboggan, or snowboard. But snow can also create problems by blocking mountain roads and preventing people from reaching their homes. There are more serious problems, too.

In spring, when the snow starts to melt, huge slabs of it sometimes break away without warning. They slide down the mountain at speeds of up to 100 mph (160 km/h) and flatten everything in their path—even trees and houses. These big snow slides are termed avalanches, and they kill people every year. If the weather warms up very quickly in spring, the runoff from melting snow can cause catastrophic floods as it cascades down narrow mountain valleys.

Factor 50

There are places in mountains where rainfall is not that great but where the sun rarely shines. Some mountain valleys and peaks lie at the height where clouds usually form. These mountain slopes remain shrouded in cloud for much of the year, and the type of plant life that grows there is called cloud forest. Cloud forest is very lush, and it is easy to understand why. Water droplets in the cloud do not fall as rain but are deposited directly on the leaves of plants. This is called horizontal precipitation or cloud stripping, because the moisture is stripped from the clouds as it condenses on the plants. Measured rainfall may be lower in cloud forest than at sites farther down the mountain, but cloud forest is always wet.

At higher altitudes still, mountains poke out above the clouds for most of the time. There, it is sunny throughout the day, and—because the air is thin and free from pollution and water vapor—the sun's radiation is very powerful. Climbers have to cover exposed parts of their bodies with strong sunscreen of up to factor 50 to prevent sunburn, even though the air may be very cold.

Clues to the Past

Mountain areas hold many clues to how Earth's climate has changed over the millennia. By measuring and dating the piles of rubble (moraines) that build up around glaciers, scientists can figure out when the climate was colder or warmer. Cores of ice are sometimes drilled out of glacial ice. These cores provide a frozen record that shows past rates of ice buildup, concentrations of gases in the atmosphere, and layers of volcanic dust—evidence of past eruptions. All these measurements can give an idea of what the weather was like in the past.

Mountains such as Grand Teton, Wyoming (below), are so tall that their peaks are often above the clouds. On such sunny mountaintops, sunglasses are vital (inset) to protect eyes from the brilliant glare of the snow.

Andes

The Andean condor soars above the Andes on 20 square feet (1.85 sq m) of wing. It is the largest bird of prey in the world, but it eats only dead or newborn animals.

Stretching the entire length of South America, the Andes include some of the highest and most barren mountains on Earth. The range is a barrier to rain—while one side of the mountains is desert, the other is cloaked in lush rain forest.

1. Quito
The Spanish built Quito on the ruins of an Inca city. Despite a major earthquake in 1917, the capital of Ecuador has the best-preserved historic center in South America.

2. Huscarán National Park
This U.N. world heritage site has deep ravines and diverse plant life, making it a place of spectacular beauty. It is a refuge for the rare spectacled bear and the Andean condor.

3. Cuzco
The old capital of the Inca civilization in what is now Peru.

4. Lake Titicaca
Legend says that the first Inca rose from this lake on the high Altiplano and went on to found the Inca empire.

5. La Paz
The capital of Bolivia, founded by the Spanish in 1548, is one of the highest cities on Earth, at 11,900 feet (3,600 m) above sea level. It has cool, sunny winters and wet summers.

6. Altiplano
Descend the eastern slopes of the high, dry plain of the Altiplano, and you will soon find yourself in rain forest.

7. Mount Llullaillaco
At an amazing 22,109 feet (6,739 m) high on the summit of this mountain, archaeologists discovered the mummified bodies of several people of the Inca civilization. This makes it the highest archaeological site on Earth.

8. Aconcagua
At 22,840 feet (6,962 m) high, Aconcagua is the highest peak in the Americas. It lies near the Chile–Argentina border.

9. Los Paraguas National Park
This park is famous for two active volcanoes and a forest of monkey puzzle trees, also called Chile pines. The park protects a patch of rare southern beech forest, too.

10. Cordillera Sarmiento
Two of the peaks in this section of the Andes—the Fickle Finger of Fate and Gremlin's Cap—are among the world's greatest challenges for mountaineers.

Inca Glory

Machu Picchu (right), near Cuzco, perches high on a massive rock, thousands of feet above the Urubamba River. It is one of the best-preserved sites surviving from the Inca civilization. Experts believe it was the mountain retreat of the emperor Pachacuti Yapanqui, who ruled the Incas between A.D. 1438 and 1471. After the Spanish conquest in the 16th century, the Incas abandoned the site. The American explorer Hiram Bingham rediscovered Machu Picchu in 1911. It may have been a religious sanctuary since there are

several religious monuments within the ruins. The windows in one of the monuments are aligned with the sun on midsummer's day and midwinter's day.

Caribbean Sea

Atlantic Ocean

Cristóbal
Colón Peak ▲
Barranquilla ●

TRINIDAD
AND TOBAGO

● Caracas

VENEZUELA

Orinoco River

PANAMA

Andes

Llanos (grassland)

● Bogotá

Georgetown ●
Paramaribo ●
Cayenne ●

COLOMBIA

Mount Roraima ▲

GUYANA

SURINAME

FRENCH
GUIANA

NORTH
AMERICA

SOUTH
AMERICA

Cordillera de
los Picachos
National Park

A m a z o n

Guiana Highlands

Savanna

ECUADOR

[1]

● Quito
▲ Cotopaxi

Pico da
Neblina ▲

Equator

Sangay
National
Park

● Guayaquil

Japurá River

Negro River

Manaus ●

Amazon River

Belém ●

Fortaleza ●

● Iquitos

f o r e s t

PERU

Ucayali River

Purus River

Madeira River

Tapajós River

Xingu River

Huascarán
Peak ▲

[2]

Huascarán
National Park

B R A Z I L

São Francisco River

Brazilian Highlands

Yerupaja ▲

Machu
Picchu

[3]

Salvador ●

Lima ●

● Cuzco

Andes

▲ Ausangate Peak

Lake
Titicaca

[4]

Mato Grosso
Plateau

Brasília ●

● La Paz

[5]

BOLIVIA

Goiânia ●

Sajama Peak ▲

Altiplano

Pantanal
(wetland)

Salar
de Uyuni

● Potosí

Paraná River

Pacific

[6]

PARAGUAY

Ocean

Atacama Desert

[7]

Mount Llullaillaco ▲

Nevado
▲ de Cachí

Gran Chaco
(shrubland)

Paraná River

▲ Galán Peak

Asunción ●

Ojos del Salado ▲
Monte Pissis ▲
Bonete Peak ▲

▲ Mount Mercedario

Aconcagua ▲

[8]

Santiago ●

▲ Mount Tupungato

Buenos Aires ●

Los Paraguas
National Park

CHILE

Andes

ARGENTINA

[9]

Patagonia

Laguna San
Rafael
National Park

N

Falkland
Islands (U.K.)

Los Glaciares
National Park

[10]

Torres del Paine
National Park

Cordillera
Sarmiento

Tierra del
Fuego

Atlantic Ocean

South
Georgia

Andes Facts

▲ The Andes run for more than 5,000 miles (8,000 km) through seven countries of South America. Only the Himalayas are higher than the highest Andes.

▲ The Andes formed when two plates of Earth's crust pushed against each other. The westward-moving South American plate started to ride over the Nazca plate about 65 million years ago, and the mountains have been growing ever since.

▲ Some parts of the Andes are desert, and some are among the wettest places on Earth.

| 0 | | 500 miles |
| 0 | 500 | km |

Mountain Plants

Plants can survive in some of the harshest mountain habitats, from smoldering volcanoes to the windswept peaks of Alaska. There are even plants clinging to life high in the Himalayas—the tallest mountains on Earth.

In winter, the Rocky Mountains look inhospitable. Deep snow carpets the ground, howling gales blast the jagged rock outcrops, and temperatures are cold enough to kill a person in minutes. That is what many mountain regions are like for much of the year. The harsh weather is too severe for human settlement, and even the hardiest animals are forced to retreat to the lower slopes. Surely no plant could survive in such an extreme environment.

Yet return to the same place in spring and you would witness an entirely different spectacle. Gone is the carpet of snow, and in its place is a flower-filled meadow. Amazingly, there are many plants that can grow and flourish in places that are covered in snow for

As soon as the snow disappears from the slopes of the Rockies in Colorado, alpine plants burst into bloom. Similar displays occur in the Alps of Europe.

most of the year and where temperatures remain freezing for months on end. In the Himalayas, plants grow on mountainsides as high as 22,000 feet (7,000 m).

Four Essentials

Plants need light, water, warmth, and nutrients to grow and to reproduce. On mountains, one or more of these essentials

is usually in short supply, at least for part of the year. Most plants get nutrients from the soil, but on rocky mountain peaks there is little or no soil. Plants receive light and warmth from sunshine, but many mountain ranges are bathed in cloud, which cuts out sunlight for much of the time. While some parts of a mountain range might receive a regular dousing of rain, others are in rain shadows and are very dry. Sometimes, even when there is water, the plants cannot use it because it is frozen.

Mountain plants face other problems, too. Many plants need insects to pollinate their flowers, but there are fewer insects on cold mountainsides than there are lower down. And the wind that blows so strongly on high mountains stops trees from growing tall.

Fact File

▲ The oldest bristlecone pines in the White Mountains of California are 4,600 years old. They were already ancient when the Mayan civilization was at its height in Central America.

▲ Microscopic plantlike organisms thrive in the boiling waters of volcanic lakes; the scalding water is so acidic that it dissolves the seams of boats.

▲ The highest forest on Earth is on the slopes of Sajama Volcano, Bolivia. It is 16,700 feet (5,100 m) above sea level, higher than any town.

▲ The *Rafflesia* plant of Mount Kinabalu, Borneo, has flowers 3 feet (1 m) wide.

Mountain Zones

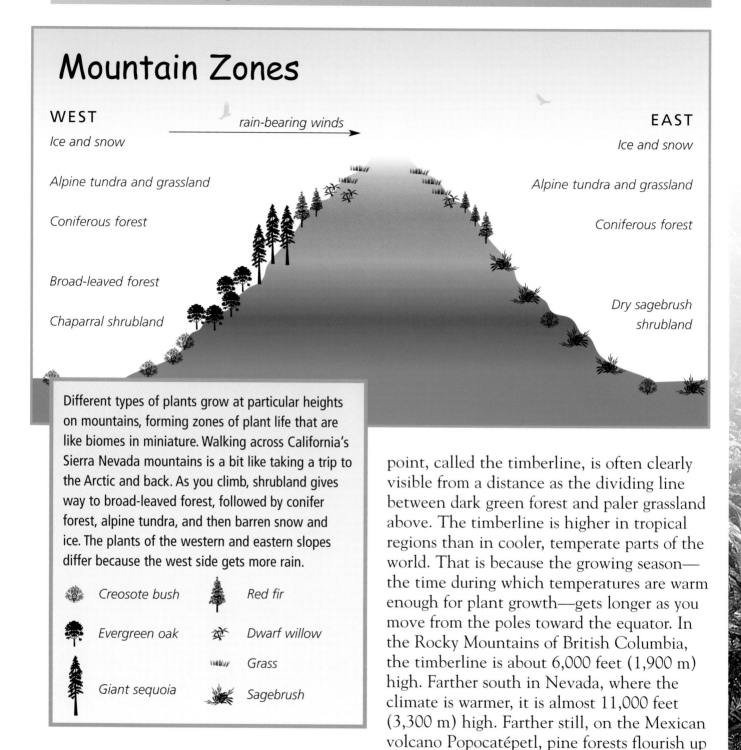

WEST — rain-bearing winds → EAST

WEST
Ice and snow

Alpine tundra and grassland

Coniferous forest

Broad-leaved forest

Chaparral shrubland

EAST
Ice and snow

Alpine tundra and grassland

Coniferous forest

Dry sagebrush shrubland

Different types of plants grow at particular heights on mountains, forming zones of plant life that are like biomes in miniature. Walking across California's Sierra Nevada mountains is a bit like taking a trip to the Arctic and back. As you climb, shrubland gives way to broad-leaved forest, followed by conifer forest, alpine tundra, and then barren snow and ice. The plants of the western and eastern slopes differ because the west side gets more rain.

Creosote bush		Red fir	
Evergreen oak		Dwarf willow	
		Grass	
Giant sequoia		Sagebrush	

The Timberline

Generally, plants become smaller the higher up a mountain you go. At lower levels, many mountain ranges are cloaked in forest. The trees tend to get smaller higher up the mountain. Above a certain height on the mountainside, no trees grow at all, and small shrubs and grasses grow in their place. This point, called the timberline, is often clearly visible from a distance as the dividing line between dark green forest and paler grassland above. The timberline is higher in tropical regions than in cooler, temperate parts of the world. That is because the growing season—the time during which temperatures are warm enough for plant growth—gets longer as you move from the poles toward the equator. In the Rocky Mountains of British Columbia, the timberline is about 6,000 feet (1,900 m) high. Farther south in Nevada, where the climate is warmer, it is almost 11,000 feet (3,300 m) high. Farther still, on the Mexican volcano Popocatépetl, pine forests flourish up to 13,000 feet (4,000 m) above sea level.

Above the timberline are zones of heath and moor before the permanent snow line. Beyond this, plant life stops and snow lies on the ground all year round. Like the timberline, the snow line gets lower the farther from the equator you go. In the very far north of North America, Europe, and

Asia, the timberline is at sea level, meaning there are no trees at all and the landscape is covered only with arctic tundra or barren expanses of snow and ice.

The Greenest Forest

On the lower slopes of tropical mountain ranges, such as the northern Andes, there is a bewildering array of big trees, forming tropical forest. The same is true in the Himalayas and many of the mountains of Southeast Asia. Under the umbrella of the tropical forest canopy is an understory of young trees and smaller plants fighting for what little light penetrates to the forest floor.

A little higher up the slopes of many of these tropical mountains is a zone that becomes engulfed in cloud from late morning

Cloud Forest and Elfin Forest

At Monteverde in Costa Rica (below), cloud forest grows between about 4,500 feet (1,370 m) and 5,000 feet (1,500 m) above sea level. An amazing variety of plants thrive in this lush jungle, including nearly 900 species of epiphytes (plants that grow on other plants) and some 450 orchids. The forest is constantly dripping, and mosses and epiphytes grow on every tree. Higher up the mountain, trees have to contend with cooler air and stronger wind—when fully grown, they reach only about 16 feet (5 m) tall. Because the trees are so small, such forest is termed elfin forest.

23

Getting a Leg-Up

Very little sunlight reaches the floor of a conifer forest, such as the forest in the Rocky Mountains. What light does penetrate the dense canopy is often blocked out by a lower layer of ferns. So how do young trees get the light they need to grow? The seeds get a leg-up on their fallen neighbors. When one of the huge conifers falls, perhaps after a storm, the top of the horizontal trunk is above the height of the ferns, so any seeds falling onto it can begin to grow and stand a reasonable chance of getting the light they need. In effect, the dead tree supports and nurtures the young saplings, so people call it a nurse log. Sometimes, trees of the same age can be seen growing in a straight line—the line of the fallen nurse log.

nearly every day. The forest that grows in such places is called cloud forest. It is an environment so lush that even the tree trunks are covered with greenery.

Small plants trying to grow in a forest of massive trees face a hard struggle. In cloud forest it is even more difficult for them since the supply of light from the sun is reduced by cloud every afternoon. Even small plants need to reach the light of the forest canopy, but how? Plants called epiphytes turn the tables on the big trees by using them as platforms. Birds eat epiphyte fruits, carry them up to their favorite perch on a large tree, then expel the seeds in their droppings. The seeds begin to grow, and their roots take hold in cracks in the tree's trunk. Epiphytic plants include orchids, and there are even cacti that grow on forest trees. Bromeliads are epiphytes whose leaves grow in a rosette shape. The rosettes collect rainwater and moisture from the cloud. Birds, frogs, lizards, and snakes drink the water, leaving their droppings before they move on. The animals' droppings provide the plants with nutrients, and the bromeliad receives far more light than it would on the forest floor.

Epiphytes get the water they need from the dampness in the clouds that sweep past daily, and they help make mountain cloud forests the greenest of all forests.

Needles for Leaves

Trees need to draw water through their roots, up their trunk, and into their leaves. In colder regions, far from the equator— and especially high on a mountain— many trees cannot draw water all year round. The water might be frozen in the ground, but even when it is liquid, the trees run the risk that it might later freeze in their leaves. Many trees simply shut down from fall to spring. Such trees have broad leaves and live

The Indian paintbrush plant is often seen clinging to rocky crevices in the mountains of Utah and Colorado. Its scarlet leaves are sometimes mistaken for flowers.

Thanks to their springy, downward-sloping branches, conifer trees shake off snow easily and suffer little damage after a heavy snowfall.

in regions that are temperate—neither very hot in summer nor very cold in winter. Every fall, the broad leaves of these trees dry out and lose their green color before falling to the ground. Thousands of people visit the Appalachian Mountains in North America each fall to witness the myriad colors—reds, yellows, oranges, and browns—that the leaves take on before they drop. Not until the following spring do the tiny green buds of a new year's growth of leaves appear.

The slopes of the Rocky Mountains are generally colder than the Appalachians. Even in summer, temperatures may drop below freezing at night. Broad-leaved trees could only stay in leaf for a small part of the year, but another type of tree—the conifer—can survive in these conditions. Conifers avoid dropping their leaves, so their leaves can continue the work of capturing sunlight and pumping water for longer. In the Rocky Mountains and other cold mountain areas, conifers such as sitka spruce, hemlock, and Douglas fir are the dominant types of trees. Their needle-shaped leaves have a protective waxy covering and only a tiny amount of freezable sap, allowing them to survive the winter freeze. The leaves do not work as efficiently as those of broad-leaved trees but do not have to be shed in fall.

Trees on the mid-slopes of the Rocky Mountains and other mountain ranges also have to deal with big falls of snow in winter. The snow's weight would break the branches of many types of trees, but mountain conifers have branches that slope downward and so let the snow slide off.

Mountain Zones

In most mountains, there are several zones of plant life, with different trees dominant at different heights. In the Rockies, spruce trees are most common at lower levels. Higher up the mountain slopes, larches and pines dominate, towering over an undergrowth of shrubs, mosses, and lichens. Higher still, the large trees are replaced by dwarf pines, birch, willow, alder, and shrubs.

Above the timberline, mountains are less hospitable. Temperatures are warm enough for plant growth during only a short part of the year. There may be more exposed rocks and less soil, strong winds blow on many days, and the ground may be covered in a blanket of snow. Although trees can't survive these conditions, many small plants occupy this zone. In some ways, the landscape above the timberline in mountains looks like the treeless wastes of the arctic tundra.

Alpine Tundra

People sometimes call the treeless zone of mountains "alpine tundra," though it is different from arctic tundra. For one thing, the Arctic gets much less rain and snow than most mountains. Its seasons are also much more severe, the long, freezing winters, when the sun hardly rises, contrasting with the short, sunny summers, when the sun hardly

Giant Sequoias

The mountains of Sequoia National Park in California are home to forests of giant sequoias, the biggest trees on Earth. Some of the greatest, most famous trees have been given names. The largest, General Sherman, is 290 feet (88 m) tall—almost one-quarter the height of New York's Empire State Building. Its trunk has a circumference of 80 feet (24 m) —the girth of a small house. If its branches, leaves, trunk, and roots were put on a giant scale, they would register more than 6,000 tons (5,400 metric tons).

Another one of these huge conifer trees, General Grant, is about 2,500 years old, and the top of its highest branch stands 267 feet (81 m) above the ground. Giant sequoias are not good trees to climb; you would need a ladder 129 feet (39 m) high to reach the lowest branch of General Grant.

Plants of the Andes

The plant life of the Andes changes along the mountain range as the warm, wet climate of the tropics merges into the cooler, more temperate south. Running along the spine of the mountains are high, windswept grasslands, known as puna. Cloud forests and rain forests flourish on the eastern slopes near the equator, where the air is misty and humid. Farther south, gnarled queñoa trees—the world's highest growing trees—survive among the dry, dusty valleys of Bolivia. Farther still, forests of southern beeches and monkey puzzles hug the slopes of the Chilean Andes, sustained by moist winds blowing off the Pacific.

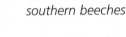

Cloud forest, featuring epiphytes

Dry mountain forest, featuring queñoa trees

Temperate, broad-leaved forest, featuring southern beeches

Grasslands

sets. Another difference is that arctic tundra has permafrost—a layer of permanently frozen ground just below the surface.

In the alpine tundra of Austria and Switzerland, plants such as the alpine snowbell lie dormant under the snow. When spring sunshine begins to penetrate the snow, the dark surface of snowbell buds absorbs the heat and helps melt the snow. This gives the plant an advantage: It becomes snow-free, and so can start to grow before its neighbors. Its beautiful mauve flowers, like those of many other plants that grow on high mountains, are large relative to the plant itself. They need to be large to attract the insects that will pollinate them, since there are fewer insects in high mountains than in the valleys and plains below. In the alpine tundra of the Rocky Mountains, there is a blanket of grasses and small plants with large, brilliant flowers in spring.

Other mountain plants are pollinated by birds, notably hummingbirds in North and South America. In Costa Rica, the bright red flowers of fuschias attract one called the magnificent hummingbird, which comes to drink nectar; in so doing it pollinates the flowers. In the Rockies, calliope hummingbirds pollinate gooseberries, currants, red columbine, and Indian paintbrush.

Cushion plants (foreground) are among the few plants that can survive high in the Chilean Andes. Their low, rounded shape gives protection from wind and cold.

Cushions and Jackets

Mountain plants risk freezing at night, even in summer. In every mountain range on Earth, there are plants with antifreeze mechanisms. In the Alps, edelweiss has a

 ## Fly Trap

Some plants reward their pollinators with an energy-rich drink of nectar, but not all plants are quite so generous. Rothschild's slipper orchid (right) grows on Mount Kinabalu in Borneo. Its flower is an amazing construction, with two long, twisted wings stretching on either side of a slipper-shaped cup. The wings have dark marks that look like a swarm of aphids. Some flies lay their eggs among aphids so that the fly grubs can use the aphids for food. The orchid's disguise clearly works since flies enter the cup to lay their eggs and, having done so, find themselves trapped inside. The only escape for the unfortunate fly is by completing a complicated maneuver to get out of the flower—which results in a bundle of pollen sticking to the fly's back.

Tough at the Top

The climate gets harsher the higher you climb up mountains, and trees find it increasingly difficult to survive. In some places, the winter wind turns trees into shrubs, creating a strange, flattened type of shrubland called krummholz (a German word that means "twisted wood"). Krummholz forms in places where plants survive the winter under a shallow blanket of snow. Although snow is cold, it acts like a quilt, protecting plants from the much colder air above. But if trees grow too tall in the summer, their branches stick out of the snow, and winter frosts soon cut them back. The result, after many years, is a high-altitude shrubland made up of stunted trees that creep along the ground.

coat of woolly hairs that protects its buds from killer frosts. In the Himalayas, *Saussurea* is completely surrounded by a protective jacket of hairs. The jacket has a hole in the top, through which insects fly to pollinate the mauve flowers. Sometimes, the insects stay the night inside the plant's shelter.

Cushion plants are protected from the cold by their shape. Their stems are packed very tightly together, forming a kind of cushion that retains warmth. The shape also preserves moisture, protecting the plants from dehydrating mountain winds. When the leaves die, they stay inside the cushion as they decay, allowing the living parts of the plant to recycle precious nutrients.

Bristlecone pines grow on the arid mountains of western North America and are thought to be the oldest trees on Earth. One of them is nearly 5,000 years old.

Bristlecone pines, which grow high in the mountains of North America's Great Basin region, also have to survive extreme conditions. From a distance these peculiar trees look dead, with gnarled and battered trunks. However, clusters of green pine needles on some of the branches show that they are alive. In fact, they are the oldest trees on Earth and can reach thousands of years old. These tough survivors live through bitterly cold winters when there is snow on the ground, and arid summers when very

A Lost World

Some of the strangest mountains are in South America. Standing thousands of feet above the rain forests of southern Venezuela are flat-topped mountains called tepuis. The vertical walls of the tepuis form some of the tallest cliffs on Earth. These mountains are nearly always shrouded in cloud, and rain falls so frequently and heavily that big puddles only rarely dry out. The rocks at the top of Roraima (above), the highest of the tepuis, look black but are really pinkish-white; their true color is hidden under a skin of algae (tiny plantlike organisms that can grow even where there is no soil). On top of the tepuis, the boggy soil contains few of the nutrients that plants need. Pitcher plants and sundews solve this problem by trapping and digesting insects, releasing extra nutrients. The plants on Venezuelan tepuis (inset) have been isolated from the forest far below for millions of years. Over this time, the plants have evolved (changed gradually) and become well suited to life in cool and very wet conditions. The mountains are home to almost 1,000 types of orchids that live nowhere else on Earth. In all, half the plants that live on these mountaintops live in no other place.

little rain falls. Such extremes mean that the pines grow very slowly—in some years they may not grow at all. Few bristlecone pines are more than 30 feet (9.5 m) tall, and many are much shorter, despite their great age.

Mountains and highlands are often plagued by strong winds that limit the growth of plants. However, in some parts of the tropics, strong winds are rare, and plants can grow relatively tall even on high mountains.

At 13,000 feet (4,000 m) in the Andes, espeletias grow taller than a basketball player. When their yellow flowers brighten the grassy mountainsides in October and November, bearded helmetcrest hummingbirds flit from flower to flower, pollinating the plants as they go. Groundsels and giant lobelias in the mountains of East Africa grow even taller, some to more than 30 feet (10 m). Like other mountain plants, they run the risk of freezing

at night—and if a plant's water supply freezes, the ice can kill it. Groundsels, espeletias, and giant lobelias protect themselves from ice in a similar way. Each year they grow a new rosette of leaves around the main stem. The leaves remain in place after they die, building up to form a protective jacket that keeps out the cold.

Groundsels (lower left, green) and lobelias (below, gray-green) grow into giants on the mountains of East Africa.

The Alps

In spring, the sunny meadows of the Alps' south-facing slopes are awash with flowers, while the musical tinkling of cowbells is everywhere. In winter, blizzards sweep across the mountains and cover their tops with snow.

Fact File

▲ Between 2 million and 10,000 years ago, glaciers filled all the major valleys of the Alps, but since the end of the last ice age they have retreated. There are still around 1,300 small glaciers.

▲ The Alps formed between 100 and 15 million years ago when landmasses from north and south moved together, gradually forcing thick layers of rock into huge folds—the mountains we see today.

▲ The Simplon Tunnel allows trains to pass under the Alps between Italy and Switzerland; the tunnel is 12.3 miles (19.8 km) long.

▲ The first people to scale the Matterhorn did so in 1865. Nowadays, about 2,000 people make the difficult climb every year.

1. Aiguilles Rouges Nature Reserve
Steep mountain slopes, which tower over several glacial lakes, are home to ibex, chamois, roe deer, and marmots.

2. Mont Blanc
Mont Blanc, whose name means "white mountain," is the highest of the Alpine peaks, at 15,771 feet (4,807 m). It is just more than half the height of Mount Everest.

3. Vallée des Merveilles
The "Valley of Marvels" has caves with thousands of rock engravings that people created during the Bronze Age, about 4,000 years ago.

4. Jungfrau
Jungfrau is one of the tallest mountains in the Alps. Massive glaciers surround it.

5. Gran Paradiso National Park
This park, together with the adjacent Vanoise National Park, is the largest protected area in western Europe. Golden eagles and eagle owls build their nests on rocky cliffs.

6. Eiger
Mountaineers consider the north face of this dramatic peak one of the toughest climbs on Earth.

7. Zermatt
One of the best skiing resorts in the world, situated under the towering Matterhorn.

8. Brenner Pass
An ancient route across the Alps, used since Roman times.

9. Lake Garda
Sheltered by the mountains, Italy's largest lake has a warm, Mediterranean climate. Scenic walking routes surround it.

10. Dolomites
This range of limestone peaks has been carved by erosion into dramatic pinnacles and gorges.

11. Salzburg
The composer Mozart (1756–1791) lived in this picturesque Austrian town.

12. Slovenia
The Slovenian Alps are riddled with limestone caves inhabited by dozens of bat species.

FRANCE

Black Forest

GERMANY

Munich

Berchtesgaden National Park

CZECH REP.

11

Salzburg

AUSTRIA

Vienna

Jura Mountains

Lake Constance

LIECHTENSTEIN

SWITZERLAND

Bern

Bavarian Alps

Innsbruck

Hohe Tauern

Graz

Aiguilles Rouges

4

6

Jungfrau ▲ ▲ Eiger

Rhaetian Alps

8

Brenner Pass

Dolomites

SLOVENIA

1

Rhône River

▲

7

10

12

Ljubljana

Matterhorn

5

Gran Paradiso

Milan

9

Lake Garda

Verona

Zagreb

2

Mont Blanc

Po Valley

Venice

CROATIA

Turin

Po River

ITALY

Genoa

Bologna

SAN MARINO

Vallée des Merveilles

3

Nice

MONACO

Mediterranean Sea

Côte d'Azur

Mediterranean Sea

0   100   miles
0   100   200 km

Corsica

EUROPE

ASIA

AFRICA

N

The breathtaking scenery around Sella Massif (left) in the Italian Alps is equally popular with skiers in winter and mountain bikers in summer.

Planning for the Future

The nutcracker (right) is a mostly brown bird, a little larger than a brown thrasher, with pale spots and a powerful beak. It lives right up to the tree line in coniferous forests in the Alps and especially favors Arolla pines. It usually lives alone or in pairs, but from July or August through fall, flocks form and spend most of the day collecting nuts and seeds, not to eat but to store. At this time, nutcrackers may fly several miles between the collecting sites and the larders where they hide the nuts and seeds. The larders are important: The birds remember where they are and raid them during winter when other food is scarce.

Mountain Animals

Mountain animals have to contend with hazardous terrain, freezing temperatures, and ferocious gales. But for those tough enough to survive, mountains are full of rich pickings, with more breeding sites than lowland areas and less competition for food.

Mountains are cooler than the lowlands around them, and the very highest peaks are usually extremely cold—even those in tropical Asia and Africa. Animals that live on mountains need ways of surviving freezing winters and bitter nights. Some have a coat of fur or feathers to keep them warm. Others can go into a deep sleep termed hibernation, allowing their body temperature to fall without doing harm to themselves. Still more animals simply move to warmer places lower down the mountain.

Guanacos live in the high grasslands of the Andes, ranging from Peru and Bolivia to the southern tip of South America. They are wild animals but are very close relatives of llamas and alpacas—the domestic animals that people of the Andes keep for wool, meat, or carrying goods. This group lives in the Torres del Paine National Park in the mountains of Chile.

Fact File

▲ The heart of a hummingbird in the Andes beats up to 1,260 times a minute.

▲ Even though they are cold-blooded animals, some iguanas live at 12,500 feet (4,000 m) in the Andes—higher than any other reptiles.

▲ A chough (a relative of the crow) once accompanied a climbing expedition up Everest to a height of 26,000 feet (7,925 m)—that's almost as high as airliners fly when crossing the Atlantic Ocean.

▲ Snow leopards have larger lungs than other big cats because they live at very high altitudes, where there is not so much oxygen in the air.

The snow leopard (below) can endure freezing Tibetan winters thanks to its luxuriously thick fur, stocky build, and bushy tail, which serves as a portable blanket.

Fur Coats

The Tibetan Plateau is so high that it is always cold, but it is too vast in area for animals to escape to warmer lowland regions. The larger animals that live there, such as snow leopards and yaks, protect themselves with a thick furry coat. The snow leopard's tail is as long as its body and much bushier than most cats' tails. When lying down, the leopard wraps the tail around its body and face for extra protection from the cold.

The snow leopard's smoky-gray color helps camouflage it against snow and rocks as it hunts for food. It preys on mammals large and small, including young yaks. But yaks have more problems to contend with than prowling snow leopards. These massive, grass-eating mammals—adult males weigh as much

A Long Sleep

To deal with the terrible winter, some animals go into a state of inactivity called hibernation. When food becomes scarce and the weather is cold, they save energy by hibernating in a warm burrow for months on end. In the coniferous forests of the Rocky Mountains, black bears hibernate for between five and seven months of the year. These magnificent animals make a good living in the forest for half the year, feeding mostly on

as a small automobile—endure temperatures that sometimes plunge to –40°F (–40°C); yaks need every hair of their shaggy coat to keep out the cold.

Feathers trap warm air in much the same way as fur does. The black-tailed ptarmigan survives even the bitterest winters in the mountains of Alaska and the Alps thanks to its insulating plumage. Like snow leopards, ptarmigan are camouflaged—in winter, their feathers turn white to blend in with the snow. Unlike snow leopards, however, their camouflage is to hide them from predators, not prey. Ptarmigan and snow leopards have something else in common: snowshoes. Snow leopards have large paws and ptarmigan have feathered feet to help them walk on snow.

Bamboo Munchers

The giant pandas that live in the bamboo forests of southwestern China have enormous appetites. Large adults munch more than 10,000 pounds (4.5 metric tons) of bamboo a year; that's the weight of four automobiles. Pandas have large, strong jaw muscles that enable them to crush the tough bamboo. It is these big jaw muscles that give pandas their rounded faces. Pandas eat young bamboo stems from November to March, old stems from April to June, and bamboo leaves from July to October. Below a height of 8,500 feet (2,600 m) most of the bamboo forest has been cut down by people, and above about 11,500 feet (3,500 m) bamboo does not grow naturally, so giant pandas have to live between these heights. That is partly why they are so rare; fewer than 1,000 now survive in the wild. The forests where they live are cold and wet for much of the year, but pandas do not hibernate. Instead, they have very thick, dense fur that is slightly oily; this fur protects them from the rain and cold.

berries, grass, roots, flowers, and nuts, but also eating carrion (dead animals), small mammals, and fish. When the weather gets colder in fall, the bears search for a good place to spend the winter months. They may select a cave or hollow tree or dig a den. Before hibernating, they feed as much as possible to put on weight—even though they use very little energy while in hibernation, they may still lose one-third of their body weight before coming out of hibernation the following April or May.

During the long winter, a black bear's heart slows down to just 8 to 12 beats a minute, and its body temperature falls by several

Above: In spring, calliope hummingbirds migrate to the Rockies from their wintering grounds in Mexico. They are North America's smallest birds—adults grow to only 3.1 inches (8 cm) and weigh the same as a quarter.

degrees. At that temperature and heart rate, the bear uses much less energy and can survive on its reserves of body fat.

Many other mountain mammals hibernate, including ground squirrels and prairie dogs. White-tailed prairie dogs live in mountain country more than 6,550 feet (2,000 m) above sea level from Montana to Arizona and New Mexico—areas where the mountains become frozen wastes in winter. The prairie dogs feed up in fall to put on a layer of fat in preparation for winter. Then, in late fall, they disappear into their burrows and go to sleep. They gradually burn off the fat while hibernating, keeping their body temperature above freezing. In spring, when they emerge once more into the outside world, they are thin and need to feed quickly on the new plant growth to get back to a healthy weight.

A few types of birds can also slow down their body to save energy. The Andes get cold at night, even close to the equator, where many hummingbirds live.

Cave Dwellers

Wherever there are mountains, there are likely to be caves as well. Though cold, dark, and wet, caves provide many animals with a refuge from enemies and inhospitable weather. Bats (right) roost in caves during the day and emerge at night to hunt insects. Bears sometimes hibernate in caves, and pumas, or mountain lions, (below) use them as lairs—gruesome feeding sites where they drag their mutilated victims.

Hummingbirds need to eat lots of sugary nectar to stay active, but they cannot feed at night when the flowers they feed on are closed. The hummingbirds' way around the problem is to go into a state called torpor—for a few hours every night, their heartbeat slows down and their body temperature drops by as much as 34°F (19°C).

Blue-spotted salamanders live in the northern Appalachian Mountains, which get very cold in winter. These amphibians survive the winter by hibernating underground but emerge early in the year when there is often still snow on the ground. They do this so they can try to be first at the breeding sites. Blue-spotted salamanders

migrate across snow and ice to reach the lakes where they breed. Even when they reach the water and dive in, they have to survive near-freezing conditions that would kill most other amphibians—some individuals even freeze. As long as there is plenty of moisture when they thaw, however, they just get up and walk away as if nothing has happened. No one knows how they survive being frozen.

Exodus

Some animals simply leave the higher slopes of mountains when the weather gets colder in fall. This way of escaping the cold is called migration and is a common strategy among birds—it is easier for them since they can fly. Migration is more difficult for mammals, reptiles, and amphibians, which have to walk. In the Himalayas, seed-eating birds such as Brandt's mountain finches eke out an existence as high as 19,700 feet (6,000 m)—higher than the highest peak in the Rocky Mountains. It makes sense for them to do so

The Resplendent Quetzal

Resplendent quetzals are brightly colored birds of mountain cloud forests in Central America. The males are famous for their incredibly long tail. Quetzals eat, among other things, the fruits of wild avocados. In fact, there are not many birds with a mouth wide enough to swallow an avocado. Flying to a favored perch, the quetzal swallows the avocado and strips the soft outer layers from the fruit before regurgitating (coughing up) the stone. Female quetzals probably use nutrients gained from eating avocado flesh to produce their eggs. So important is this fruit to quetzals that the birds often nest close to an avocado tree.

spend much of their lives in underground burrows. They sleep in their burrows at night, hibernate in them in winter, and raise their young in them. At night, pikas in the Himalayas and Tibet share their burrows with birds named snowfinches, which also need shelter. The birds sometimes take more advantage of the hospitality of their pika

because there is little competition for food at that height. They can breed and rear their young in the summer when food is plentiful, then migrate to lower levels of the mountain in winter when the upper levels are blanketed with snow. Birds called longspurs make a similar annual migration in the Rocky Mountains.

Instead of hibernating, pikas collect plants in autumn and dry them to make hay, providing a store of food to last through the winter.

Gimme Shelter

Mountains are often very windy places, and the chilling effect of wind can make animals lose heat quickly. That is not a problem for an active adult animal, but it can kill an inactive or young animal. Ground squirrels and prairie dogs in the Rocky Mountains, pikas on the Tibetan plateau, and giant mole-rats in the Bale Mountains of Ethiopia all

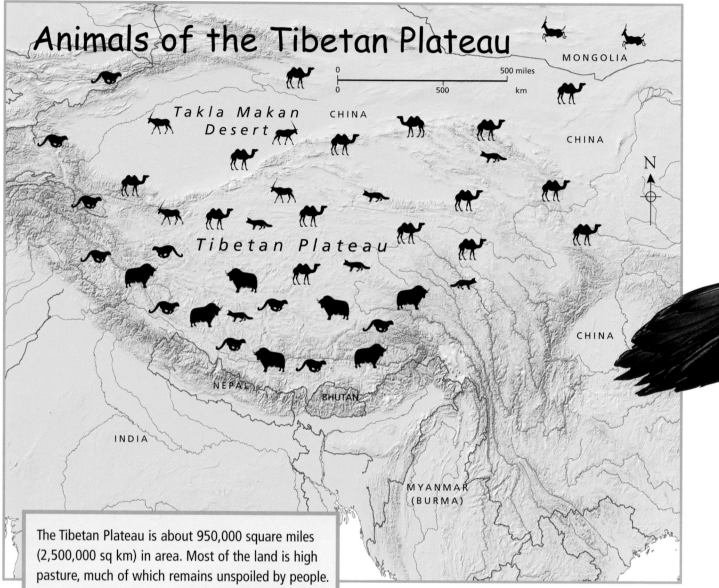

Animals of the Tibetan Plateau

The Tibetan Plateau is about 950,000 square miles (2,500,000 sq km) in area. Most of the land is high pasture, much of which remains unspoiled by people. The north especially is a desolate, arid wilderness, where only wild animals and occasional nomads set foot. Rare Tibetan antelopes and wild Bactrian (two-humped) camels live on the sparse grass of the plateau and deserts to the north. They share the area with snow leopards, Tibetan foxes, Mongolian gazelles, and yaks—both wild and domestic.

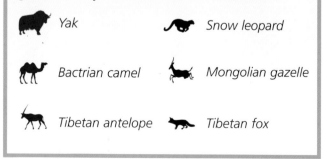

Yak

Bactrian camel

Tibetan antelope

Snow leopard

Mongolian gazelle

Tibetan fox

hosts and build their nests in the burrows. The mammals and birds seem to live together without any problems.

Visitors Move In

Mountains are places of extremes. In winter, the mountain landscape may be covered in snow, and food may be very difficult to find. Even if not snowbound, there is little food in winter. Yaks and other plant-eating animals on the Tibetan Plateau are forced to exist on a diet of dry, dead plants for most of the year—green shoots are available for only three months or so.

A bone clutched in its claws, a lammergeier soars over the Drakensberg mountains of South Africa. These huge vultures nest on rocky ledges among mountains. They feed on carrion (dead animals) and break open bones by dropping them from a great height.

In summer, however, when the snow has gone, the mountainside may be clothed in flowers, with their attendant insects. For those animals able to take advantage of it, the mountain then becomes a place of plenty, even if only for a few short months. Visitors make incredible journeys to take advantage of the insects and other invertebrates that appear on the slopes. The northern wheatear is an insect-eating bird that spends the winter months in Africa. Each spring, many thousands of these birds desert their winter quarters. Some fly thousands of miles to the high-altitude meadows of the Alps, but others go much farther: to the mountains and highlands of Scandinavia, Central Asia, Greenland, and even Alaska. The birds nest in their summer home, and in late summer they and their offspring make the epic return journey—across land and sea—to Africa.

Sure-footed and agile, mountain goats can negotiate even the most treacherous cliffs and glaciers with effortless confidence. They shed their woolly coats each summer.

Bone Breakers

Vultures and other birds of prey soar high on spirals of warm air termed thermals. From their vantage points high in the sky, the birds can search for dead animals lying on the mountainside. One type of vulture, found in the mountains of Europe, Asia, and Africa, has an ingenious method of increasing its intake of protein: The lammergeier picks up a bone from a dead animal and then soars very high. Once high enough, it drops the bone, which usually breaks into pieces on the rocks below. The lammergeier then flies down to feed on the nutritious marrow. Although it is a large and impressive vulture, it is dwarfed by a vulture in the Andes. The Andean condor is the largest flying bird on Earth; its wings span the length of a small automobile. Like the lammergeier, it feeds mostly on dead animals.

Gentle Giants

High in the mountains of southwestern Uganda, Rwanda, and the Democratic Republic of the Congo live about 600 of our closest relatives—mountain gorillas. In the past, gorillas gained a fearsome reputation because of their size and our lack of understanding about how they live. Films such as King Kong encouraged the view that gorillas are aggressive monsters, liable to attack and kill people without provocation—but nothing could be further from the truth. They are, in fact, gentle giants.

Mountain gorillas certainly make an impressive sight. A fully grown male has a wide chest and powerful legs and arms. He may weigh 400 pounds (180 kg)—two and a half times the average weight of a man—but a female may be half this size. Despite their appearance, gorillas are shy animals that feed on plants and fruit, supplemented with termites. They have few enemies except people. Relentless hunting and destruction of mountain forests by people have brought mountain gorillas to the brink of extinction. If the mountain forests disappear, the last gorillas will disappear with them.

Mountaineers

Mountain animals have to get around in difficult terrain. Two animals that live in the Alps and Pyrenees of Europe—the ibex (right) and chamois—are masters of the art of climbing and jumping. Both species make leaping from rock to rock and climbing near-vertical slopes look easy. In the summer, ibex feed in meadows and on rocky ground up to 10,000 feet (3,050 m) high in the Alps, but they descend to lower levels when the weather turns colder in fall.

Timed for Success

Animals are most vulnerable to cold and hunger when they are young. Mammals and birds that breed in harsh mountain environments time their courtship and breeding so that the young are born or hatch when conditions are most favorable. Every June, female Tibetan antelopes hurry north across the very high Tibetan Plateau, following an ancient migration route north to sites in the Kunlun Mountains (no one knows exactly where) to give birth. Two months later, the mothers return on their long trek south to meet with the male antelopes, this time accompanied by their offspring. Even in midsummer, blizzards often sweep across the vast plateau; when they do, many young antelopes die.

Blizzards are not the only hazard the antelopes have to contend with. Predatory snow leopards with their own young to feed prowl the plateau in search of weak antelope to prey upon. Female snow leopards give birth to two or three cubs in late March or April. The cubs remain dependent on their mother for almost two years, and during this time the mother is constantly on the lookout for a meal. A good-sized kill—an ibex or blue sheep, for example—will last for several days.

Right: Mountain butterflies emerge from pupae in summer to feed in the flower-filled alpine meadows.

East African Highlands

From snow-clad mountains to lush cloud forests and boggy heaths, East Africa's highlands follow the massive rift valleys that stretch north from the Zambezi River to the Jordan River, past the Red Sea.

Breaking Apart

The African continent is slowly breaking apart. Most mountains in East Africa are either volcanoes or great masses of rock forced up by volcanic activity. They run either side of two huge valleys—the Great Rift Valleys—that plow through the region. The rift valleys and mountains were created by the same gigantic processes deep within Earth. About 7 million years ago, two plates of Earth's crust, the African and the Arabian plate, began to separate. Lava (molten rock) from deep inside Earth pushed to the surface, creating volcanoes. Where the lava could not find an outlet, the pressure it exerted pushed the Earth's surface higher. The pressure created more faults (cracks) and more seepages of lava. These processes are still at work today.

1. Lalibela
Several ancient Christian churches were carved from the rock in and around this old Ethiopian capital.

2. Addis Ababa
Four million people live in this, the modern capital of Ethiopia. The city's National Museum houses Lucy, the skeleton of a very early human ancestor.

3. Bale Mountains National Park
The moorland in this national park is the highest in Africa and is home to rare animals and birds, including the Ethiopian wolf and the giant mole-rat.

4. Ruwenzori Range
The highest mountains in Africa that are not of volcanic origin. They are sometimes named the "mountains of the moon" for their snow-covered peaks.

5. Mount Elgon
The snows and streams of this mountain provide a vital source of water for several million people in eastern Uganda and western Kenya.

6. Mount Kenya
A cool sanctuary in a very hot region, this extinct volcano has 11 permanent glaciers and many types of plants found nowhere else on Earth.

7. Lake Victoria
Uganda, Kenya, and Tanzania share the waters of the largest lake in Africa.

8. Nairobi
The arrival of the Kenya–Uganda railroad a century ago triggered the birth of Nairobi, one of Africa's biggest cities.

9. Mount Kilimanjaro
Three extinct volcanoes combine to form this famous snowcapped mountain. Carved stone bowls found on its slopes are 2,000 years old.

10. Bwindi Impenetrable Forest
Lush, dense, and often very rainy, this forest is home to ten species of primates, including mountain gorillas and chattering colobus monkeys.

11. Lake Tanganyika
Mountain slopes rise steeply from the lake's western shore. On the east are the rugged hills of Gombe Stream National Park, where scientists study chimpanzees. The lake's fish are important food for local people.

Gelada baboons live only in remote parts of the Ethiopian Highlands, where they perch on inaccessible cliffs.

Nile River

500 miles

0
500
km

N

Khartoum

SUDAN

SAUDI
ARABIA

ERITREA

Red Sea

YEMEN

YEMEN

DJIBOUTI

Asmara

Denakil Desert

Ras Dashen

Simien Mtns.

Abune
Yosef

Guna Terera

Lalibela

1

Ethiopian Highlands

Blue Nile River

Addis
Ababa

2

Mendebo Mtns.

Bada

K'ech'a Terara

Batu

Guge

ETHIOPIA

Bale Mountains
National Park

3

White Nile River

Sudd
(wetland)

Lake Turkana

Great Rift Valley

SOMALIA

DEMOCRATIC REPUBLIC OF THE CONGO

UGANDA

4

Ruwenzori Range

Mount
Stanley

5

Mount Elgon

Kampala

KENYA

Mount Kenya

Equator

7

6

Lake
Victoria

Nairobi

RWANDA

8

Kilimanjaro

9

BURUNDI

Bwindi
Impenetrable
Forest

10

Ngorongoro
Crater

Mombasa

Pemba Island

Gombe Stream
National Park

Zanzibar Island

Lake Tanganyika

Mahale Mtns.
National Park

Dodoma

Dar es Salaam

Rubeho
Mtns.

Mbizi
Mtns.

11

TANZANIA

Kipengere Range

Livingstone Mts.

Indian Ocean

Lake
Mweru

COMOROS

ZAMBIA

Lake
Bangweulu

Lake Malawi

MALAWI

Luangwa River

Lilongwe

MOZAMBIQUE

Lusaka

Zambezi River

ASIA

AFRICA

Harare

ZIMBABWE

Bulawayo

SOUTH
AFRICA

MADAGASCAR

Fact File

▲ The seven highest mountains in Africa are in the
East African highlands.

▲ The cool, well-watered East African highlands
include some important areas for agriculture. Farmers
grow coffee, sugarcane, and cotton in Ethiopia, and
tea, coffee, and flowers in Kenya and Uganda.

▲ Unlike mountainous regions elsewhere, some
highland areas of Ethiopia are among the most
densely populated places in Africa.

*Strange plants such as the giant lobelia (below)
grow on the high moorland of the Simien
Mountains in northern Ethiopia.*

People and Mountains

Mountains and highlands can be inhospitable or even deadly places, yet people have eked out a living in them for thousands of years. Today, they are a magnet for thrill-seekers, from mountaineers to skiers.

Relatively few people live in Earth's mountains and highlands, and it is easy to see why. The cold, windy, and often wet climate can make life uncomfortable; the steep, rocky slopes are difficult to farm; and the short growing season makes it hard to raise crops. Because of the rugged terrain and the difficulty of building roads, mountain locations are also tricky to reach and slow to travel through. These are some of the reasons that have kept people from living in many highland regions. Not all mountains are freezing, rocky wastelands, though. In the tropics, some people prefer the more open highland environments, where they don't have to hack their way through impenetrable, insect-infested jungle to get around. And while tropical lowlands are often hot and sticky, the highlands have a cooler, more comfortable climate. The relatively rich, fertile soils of the Ethiopian Highlands have been that country's bread basket for many centuries. Likewise, the fertile volcanic soils of the central Mexican highlands were the cradle for the great Aztec civilization.

Early Civilizations

In most mountain ranges, people left few signs of their activities until the last few centuries. The high central plateau region of Mexico is an exception. Archaeologists estimate that some tools and other artifacts discovered at the ancient settlements of Tlapacoya and Tepexpan were crafted by people at least 20,000 years ago. Why were they drawn to this highland area?

The area is a relatively sheltered basin, surrounded by even higher mountains. Some of the mountains were—and still are—active volcanoes, spewing lava into the basin. The soil that forms after the lava is weathered by sunshine, rain, and frost is rich in minerals, helping the growth of plants. The earliest human settlers discovered that they could grow fruit and vegetables in this rich soil. Over time, they modified their environment, cutting down trees to build houses, to collect firewood, or, later, to clear the land for crops. The remains of an early town at Tlapacoya date from between 3,500 and 2,500 years ago, and people built other settlements around the shores of lakes in the area. No one knows exactly how these people lived, but we do know that they grew crops, fished the lakes, and hunted mammals and birds in the forests.

The people of the area around present-day Mexico City prospered. The city of Teotihuacan was home to many thousands of people at its height in A.D. 650. People from other areas heard about the region's prosperity and moved in, and about 600 years ago a tribe called the Aztecs arrived and started building what was to become the city of Tenochtitlan on an island in Lake Texcoco. The Aztecs drained part of the lake to build their city and built floating platforms

Forbidding Lands

Although people have lived in mountain regions for thousands of years, mountains have often proved inaccessible to outsiders. Early explorers wrote of the problems of mountain travel. It took the 13th-century explorer Marco Polo several years to cross the mountains of Iran, Afghanistan, and Pakistan, for example. Even today, mountains are a barrier to travel, often separating different countries or regions with very different cultures. High in the Himalayas of Nepal (below), travel by car is impossible because the slopes are too steep for permanent roads. Locals travel by foot and measure distances in hours or days rather than miles or kilometers.

A Mountain Empire

The Inca were an ethnic group originally based around what is now the town of Cuzco in Peru. By the time of the Spanish conquest of South America in 1532, the Inca ruled more than 12 million people of dozens of different cultures and speaking at least twenty languages. Inca conquerors brought many benefits to their subjects, including fine textiles and agricultural know-how. They allowed local leaders to remain in power but took their sons back to Cuzco for training. At its peak, the Inca civilization straddled the Andes along most of their length, from Chile in the south to Colombia in the north—roughly the distance between New York and Los Angeles.

of reeds and leaves called *chinampas* on which to grow crops. The farmers and fishers of Tenochtitlan traded fruit, vegetables, animal skins, and fish with neighbors who traveled to the city's market. Different parts of Tenochtitlan were connected by a network of canals, and the city eventually boasted palaces, temples, a zoo, plazas, markets, and aqueducts carrying water from springs. Within 200 years, these highland people were able to conquer the neighboring peoples for hundreds of miles around. When the Spanish conquistadores arrived in the 16th century, they were amazed at Tenochtitlan's sophistication and compared it with the greatest cities in Europe.

Wise Farmers

Wherever people occupied mountains, they changed their surroundings. They cut down trees, plowed slopes for crops, and built towns and villages. The Inca people of South America were expert farmers. They realized that if they chopped down trees the rain would wash away the soil, so they built steps called terraces on the steep slopes. The

terraces allowed them to farm while keeping the rich soil from washing away downhill. Mountain people in parts of Southeast Asia also construct terraces that follow the contours of the land.

Many crops grow better in cool mountain climates than in lowland regions. Most of the

world's tea comes from highlands in East Africa, South Asia, and China; coffee is grown in the mountains of Central and South America. Unfortunately, people have not always followed the good agricultural practices of the Incas in mountain regions. Slash-and-burn farmers grow crops on tropical mountains in a way that can damage the natural environment. First they cut down and burn all the wild trees to clear the ground. Then they plant a quick-growing crop. With no trees to protect the ground, rain soon washes away the fertile soil. After a few years, the land is useless for crops and the farmers move elsewhere, repeating the cycle and destroying another patch of forest.

Versatile Mountain Beasts

The people who live on the Tibetan Plateau have shared their highland home for thousands of years with animals called yaks. These hardy mammals, which look like very hairy cows, play an important part in the lives of Tibetan people. Yaks live at much higher altitudes than other mammals—up to 20,000 feet (6,000 m) above sea level. Their shaggy fur coat protects them from freezing temperatures, and their large lungs help them breathe the thin mountain air. Yaks are strong and sure-footed. An adult yak can carry the weight of two men over rough terrain. Since yaks can walk dangerous, snow-covered Himalayan trails, people use them to transport loads. Yaks plow fields and provide milk, butter, and wool. People use yak dung for making fires, yak hair for ropes, sacks, blankets, and tents, and yak bone to craft tools. Nothing is wasted.

Riches from the Depths

The same mountains that can make life so hard can also provide unimaginable wealth. People have mined and quarried the rocks of highland regions for thousands of years for their precious or semiprecious minerals. Gold, silver, tin, copper, and lead are some of the metals that are found in mountains, while there may also be diamonds and other gemstones. Mining is not just a recent activity. The mines of the Potosi area in

The dusty village of Kagbeni is deep in the Nepali Himalayas, 9,200 feet (2,800 m) above sea level and close to the Chinese border. This is where the humid climate of the southern Himalayas gives way to the much drier, desertlike climate of the Tibetan Plateau. The people who live here use meltwater streams to irrigate crops of barley.

The Inca of South America built terraces (stepped fields) on hillsides in the Andes. Terraces are still used in many parts of the world to grow crops, from rice to grapes.

Bolivia were first dug into the mountainside almost 500 years ago. Men, women, and children worked the mines for silver and tin. So great were the riches extracted from them that Potosi—the highest city on Earth, at 13,000 feet (4,000 m) above sea level— became South America's biggest city in the 18th century. With mining came the need for wood to support the shafts and tunnels, so people cleared forests for timber. The miners needed houses, so they quarried stone and felled more trees. They built roads to connect their mines to the coast, and developed other industries nearby to take advantage of the raw materials. All of these activities transformed the mountain environment.

Since the 18th century, iron has been one of the most important raw materials for the people of North America and Europe. Powerful furnaces were needed to release iron from rock, and such furnaces were usually fueled by charcoal (partially burned wood).

The people of Tibet domesticated yaks many centuries ago. Domestic yaks are much smaller than their wild cousins, which can reach more than 6 feet (1.8 m) tall at the shoulder.

When iron was discovered in the Appalachian Mountains of the United States, people began cutting down the trees to make charcoal. Vast areas of forest were cleared in the process. Today, the need for charcoal has all but disappeared, and much of the Appalachian forest has grown back.

Commercial logging is now the main reason for felling trees in North American mountains. Thousands of acres of softwood (conifer) trees are felled each year in parts of the Rocky Mountains and elsewhere. The timber is then used for building houses or for making furniture or paper. Many people fear that the logging industry cannot continue at such a pace because the forests will dwindle and disappear. In many parts of the world, new trees are planted to replace those cut down, but that is not true everywhere, and there are still conflicts between the interests of loggers and conservationists.

Thirst Quenching

In very high mountains there is little or no history of human settlement, and this part of the mountain biome remains largely unspoiled. However, in recent years things have changed. Freshwater is in increasing demand as the world's population grows—people need water for drinking, cooking, and washing. Water irrigates parched land so the land can support crops, and water drives the turbines of power stations. Many mountain rivers have been dammed to create reservoirs that provide water for cities and industry and power to drive the turbines in hydroelectric

Steep-sided valleys provide ideal sites for dams. Such dams create reservoirs of freshwater as well as generating electricity, but they can damage the environment. Hoover Dam (below) submerged part of the Colorado River valley, forming Lake Mead—a reservoir 250 square miles (650 sq km) in area.

plants. Dams also help control rivers that flood seasonally. The once-wild Colorado River, for example, used to flood in spring as it swelled with meltwater from the Rockies. Hoover Dam and Glen Canyon Dam stopped the floods and created scenic lakes that are now popular with hikers, swimmers, sightseers, and boaters. But such dams and lakes come at a cost—they drown the valleys upstream, killing the animals and plants that once lived there.

Scaling the Heights

In recent times people have begun to exploit mountains more for their leisure potential than for their mineral wealth. Skiing, climbing, hiking, mountain biking, and ecotourism attract millions of people to mountains every year and provide work for many

🥾 *hiking*

⛷ *skiing*

🧗 *climbing*

🏂 *snowboarding*

NORTH AMERICA

SOUTH AMERICA

Mountain Sports

To mountain sports enthusiasts, the world's great mountain ranges are an enormous playground. Skiers and snowboarders seek snow-sure slopes in winter, preferring cooler climates and countries with high-tech resorts. Hikers and mountaineers often look further afield for a challenge, tackling mountains throughout the tropics. The world's mountains also attract hang-gliders, rappellers, tobogganists, and countless day-trippers.

Into Thin Air

The air at the top of a mountain is thinner than air at sea level, which can make it difficult to breathe properly. Anyone who climbs a tall mountain quickly risks suffering from a condition called altitude sickness, caused by lack of oxygen in the body.

Altitude sickness is most common above 8,000 feet (2,440 m). At first it causes headache, giddiness, and heavy breathing, but these problems clear up if the climber descends or waits a few days while the body adapts. However, some climbers are not so sensible. If people with early symptoms of altitude sickness keep climbing, the condition can make their lungs or brain fill with fluid. Unless they descend, they can fall into a coma and die within hours.

Mountaineers can protect themselves from altitude sickness by carrying an oxygen supply. Even so, heights above 25,000 feet (7,600 m) are known as the "death zone," and all mountaineers who climb this high get so out of breath that they have to move very slowly.

A climber takes a well-earned rest on the summit of Mount Whitney, California.

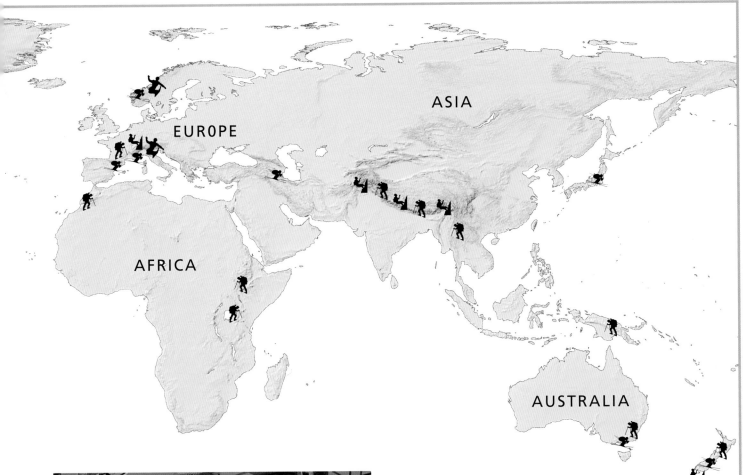

ASIA

EUROPE

AFRICA

AUSTRALIA

more. Skiing is particularly big in the Alps, Rockies, and northern Appalachians, while climbers flock to any range with challenging rock faces. The Himalayas are renowned for their hiking, while the birds and other wildlife of the cloud forests of Costa Rica attract thousands of ecotourists every year. Badly managed, tourism can be very damaging to natural environments. However, with the right controls, the expanding leisure industry perhaps offers the best hope for protecting the last of the world's great mountain wildernesses.

Skiers take the chairlift to Squaw Peak, California, near Lake Tahoe. Lifts are also used by sightseers in summer, but some hikers consider them an eyesore.

53

Himalayas

The soaring peaks of Earth's tallest mountains formed as the Indian landmass collided with Asia, squashing and folding the land over millions of years. The mountains are still rising today.

Fact File

▲ The Himalayas include our planet's 15 highest mountains. Mount Everest is the tallest.

▲ Every year, Mount Everest grows in height by about the thickness of two cents.

▲ Tibetan people carry out "sky burials" when someone dies. Instead of putting the body in a grave, they chop it up and let vultures eat it.

Because It's There

People have always been fascinated by the challenge of climbing mountains. Mount Everest (below), the greatest of them all, is no exception. The first seven recorded attempts on the summit ended in failure, but in 1953, Edmund Hillary and Tenzing Norgay reached the top. Since then, more than 600 climbers from 20 countries have scaled Everest. Milestone climbs include those by Junko Tabei, the first woman, in 1975; Tom Whittaker (right), the first disabled climber, who climbed Everest in 1998 with an artificial leg; and

Reinhold Messner, the first to reach the summit without oxygen. Having climbed it once, would anyone want to go back to the top? Yes! A sherpa (guide) named Ang Rita has performed the climb 10 times so far. There have been many sad endings, too: at least 100 climbers have died in pursuit of the ultimate climbing glory.

Glacier

Mount Nuptse

route to summit

Mount Lhotse

Mount Everest

Glacier

Glacier

1. Valley of Flowers National Park
Hundreds of different wildflower species carpet the meadows of this small reserve.

2. Govind National Park
The rare snow leopard lives in the higher parts of this Indian Himalayan reserve.

3. Source of the Ganges
The Ganges River is sacred to Hindus. It flows through India and Bangladesh, where it empties into the Indian Ocean.

4. Karakoram Pass
One of the highest trade routes on Earth, this pass links India with China.

5. Kangrinboqe Feng
Pilgrims from Nepal, India, and China consider this round-topped mountain sacred.

6. Kathmandu
Nepal's capital lies in the foothills of the Himalayas. Its temples are overrun by monkeys, which the local people consider sacred.

7. Mount Everest
The highest mountain on Earth at 29,035 feet (8,850 m).

8. Tibetan Plateau
A huge area of highlands to the north of the Himalayas. The mountains block rain-bearing wind, giving Tibet a much drier climate than India.

9. Chang Tang Reserve
Antelope, yaks, and pikas live in this vast reserve, which is dotted with salt lakes. Chang Tang is bigger than Arizona.

10. Siling Co
Fed by glacial meltwater, this is the biggest lake on the Tibetan plateau. Tibet has the highest concentration of lakes in China.

11. Lhasa
The capital of Tibet has some beautiful buildings, including magnificent Potala Palace, with more than 1,000 rooms.

12. Source of the Chang
From its source in Tibet, the Chang (Yangtze) River flows through China to Shanghai.

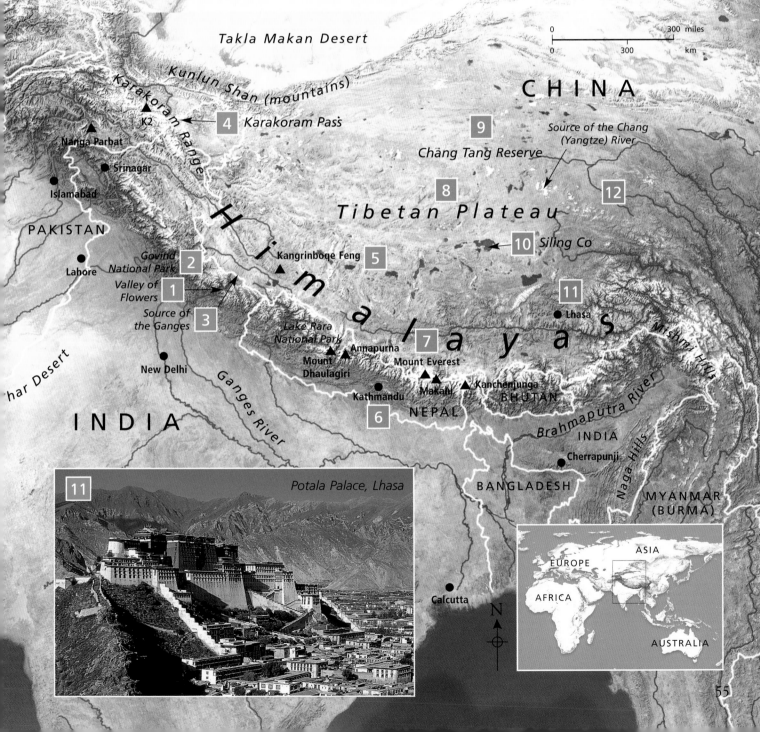

Takla Makan Desert

Kunlun Shan (mountains)

CHINA

Karakoram Range

K2

Nanga Parbat

Srinagar

Islamabad

PAKISTAN

Lahore

Thar Desert

New Delhi

INDIA

Ganges River

4 Karakoram Pass

9

Source of the Chang (Yangtze) River

Chang Tang Reserve

8

Tibetan Plateau

12

10 Siling Co

Govind National Park 2

Valley of Flowers 1

Source of the Ganges 3

Kangrinboqe Feng 5

Lake Rara National Park

Mount Dhaulagiri

Annapurna

7

Mount Everest

Makalu

Kanchenjunga

11

Lhasa

Mishmi Hills

Himalayas

Kathmandu 6

NEPAL

BHUTAN

INDIA

Cherrapunji

Brahmaputra River

Naga Hills

BANGLADESH

MYANMAR (BURMA)

Calcutta

N

11 Potala Palace, Lhasa

EUROPE

ASIA

AFRICA

AUSTRALIA

The Future

The natural processes that shape mountains never stop, so they are ever-changing and sometimes violent places.

Forces from deep in Earth's crust are continually altering the shape of mountains, although the change is usually gradual. Volcanoes and earthquakes can reshape mountains in sudden, spectacular, and sometimes deadly fashion. Wind, rain, frost, and rivers mold mountains over thousands of years. And human activities are now accelerating some of these changes.

It is probably more difficult to predict how mountains might change in the future than how other biomes will. The raging torrents that bring meltwater down mountainsides can swell so suddenly in the spring that trees, boulders, and rocky cliffs are swept away in their path. Huge avalanches of snow can smash their way through acres of forest. A volcano can blow its top with such ferocity that it is a completely different shape after the eruption. There are more earthquakes in mountain regions than in any of Earth's other biomes, and a large earthquake can make a

The eruption of Mount St. Helens obliterated the top 1,300 feet (400 m) of the mountain and blasted ash and dust 15 miles (24 km) into the sky.

mountain higher or lower than it was before. Scientists try to work out when these gargantuan events will take place, but most of them cannot yet be predicted.

There She Blows!

Mount St. Helens in Washington erupted on May 18, 1980, after an earthquake caused half the side of the mountain to collapse. This, in turn, sparked an upsurge of molten rock from deep underground, and huge quantities of rock and ash were blasted out of the top of the volcano. If all the trees blown down in the blast were put end to end, they would stretch from Earth to the Moon and back twice—that's enough trees to build 300,000 two-bedroom houses. Within 15 minutes, the column of smoke from the crater reached 80,000 feet (24,000 m) tall. Two-thirds of a cubic mile of rock collapsed

Below: Volcanoes can build mountains as well as destroy them. Hawaii's volcanoes rose gradually from the seabed as lava from successive eruptions piled up.

Northern Spotted Owl

One of the bitterest disputes between conservationists and loggers took place in the mountains of the Pacific Northwest, home of the northern spotted owl. This rare bird lives only in old, undisturbed forest. The loggers argued that employment depended on felling the trees, while the conservationists replied that the birds would become extinct if the chain saws advanced any farther. People organized protest rallies, others filed lawsuits, and scientists tried to come up with a solution to the problem. Eventually, the federal government devised a way of conserving the old-growth forests of western Oregon, Washington, and northern California. The loggers can continue their work as long as they do not get too close to the owls' habitat. So, for the time being, the future of both the timber companies and the owls is secure.

in a landslide that traveled at up to 150 mph (240 km/h) across 23 square miles (60 sq km). Human casualties numbered only 57, but millions of mammals, fish, birds, and other animals were killed in a few minutes.

Lumberjacks and Loggers

Not all changes that happen in mountains are natural. For centuries, people have cut down the forests that grow on lower slopes of mountain ranges. If people cut down the forest slowly or in a planned way, the forest has time to regenerate; young trees grow in the gaps created by the felling, and little long-term damage is done. However, in many highland areas, logging companies clear entire mountainsides of trees, sometimes with catastrophic results. Unless this wholesale forest clearance stops, more soil will wash away, and slopes that previously supported a wide range of plants and animals will become almost barren rocky wastes.

Mining and quarrying can create huge, ugly scars in mountains. People use strip mines when the minerals they need are near the ground surface. Instead of tunneling deep into a mountain, they strip the soil away and dig a wide hole on the surface. This type of mining removes all animals and plants from the mined area. Even after workers abandon a strip mine or a quarry, it is many years before plants grow again on the rocky

cliffs. And for every new quarry that is blasted, workers build a road leading to it, houses for the miners, and possibly other developments that damage the environment.

Global Warning

Even greater will be the changes triggered by global warming. When automobiles and factories burn fossil fuels, such as coal and gasoline, they release carbon dioxide into the atmosphere. This gas stops heat from escaping from Earth into space, so the planet gradually gets warmer. Some signs of global warming have been clear in mountains for several decades. For example, glaciers are getting smaller, and the skiing season in the

Bingham Canyon Mine near Salt Lake City, Utah, is the largest strip copper mine in the world. Mines such as this can transform mountain landscapes.

On the Slide

When loggers strip forest from a mountainside, trees' roots no longer hold the soil in place. Within a few years, rain washes the soil away and begins to erode the ground. The badly eroded valleys of Madagascar (above) illustrate this process. Deforested slopes can also become dangerously unstable after rain. The torrential rain that accompanied Hurricane Mitch's passage across Central America in 1998 resulted in hundreds of mud slides, killing thousands of people and making many homeless. If the region's forests had not been cut down, there would have been fewer casualties and less damage.

Alps is getting shorter. In the future, forests might creep higher up mountains and invade the wildflower meadows of the upper slopes. Strangely, global warming could make some mountain areas colder by disrupting global weather systems; no one yet fully understands how global warming will change Earth's climate in the years to come.

Pressure and Pleasure

As Earth's population grows, there will be more pressure to exploit mountains for their minerals, timber, and water. The demand for freshwater for drinking, irrigation, and industry will certainly increase. Highlands can provide the biggest, easily accessible source of this water. Many mountain rivers have already been dammed to create reservoirs, the water from which is pumped to lowland cities and farmland. People are likely to dam more rivers in the future, but for every reservoir that is built, a valley and the wildlife within it are flooded, while the river itself undergoes profound changes that affect its wildlife. Many young people have left their remote, small communities in parts of the Andes and Himalayas to seek work in big cities in the lowlands. The older people tend to

? Moving Mountains

The Himalayan range, which stretches 1,300 miles (2,100 km) from Pakistan (shown here) in the west to Bhutan in the east, is the site of the greatest collision on Earth. The Indian plate of Earth's crust is crunching into the Eurasian plate at a rate of almost 1 inch (2.5 cm) each year. Eventually the rock must fracture, and when it does there will be a series of enormous earthquakes. Experts believe these earthquakes are overdue, and that when they do happen they could shatter large dams, releasing vast reservoirs of water. Fifty million people are at risk in a zone running along the southern side of the Himalayas.

stay in the mountains, so the mountain population ages. Without the next generation of young people, centuries-old traditions and cultures will gradually become extinct.

One way to breathe new life into these areas is to encourage people to visit and enjoy them. Mountains are great places for outdoor pursuits. Skiing, snowboarding, hiking, fishing, rock-climbing, and watching wildlife are a few of the activities that attract millions of people to mountains every year. Though snowmobiles, ATVs, and campsite litter all take their toll on the environment,

tourism does much less damage than mining or logging. By encouraging leisure pastimes and ecotourism in mountain areas, governments help to preserve the future of these wild areas—and at the same time assist the people who live there.

Parks and Preserves

Denali National Park in Alaska preserves spectacular landscapes, including Mount McKinley, the highest mountain in North America. It is also home to grizzly bears, beavers, moose, and wolves. On the other side of the world, the Lake Rara National Park in Nepal was established in 1975. It is one of the least-visited preserves on Earth, but it gives full protection to its animals and plants. Rara nestles between steep forested ridges and snowy Himalayan peaks. The lake at its center is the largest in Nepal, while the forests of spruce, juniper, and pine support Himalayan black bears, red pandas, and a varied selection of birds.

Every year sees more preserves established and more of the world's mountain biomes given protection. Sometimes the interests of leisure and industry are at odds. What happens, for example, if prospectors discover oil within a national park? Should they be allowed to drill for it? There are no easy answers, but at least nowadays people are more aware of the importance of the plants and wild animals of our mountains and highlands, so they have a better chance of survival.

Glossary

air pressure: A force exerted by air in the atmosphere as it weighs down on Earth. Air pressure decreases with increasing altitude.

alpine tundra: A treeless zone similar to arctic tundra above the timberline on mountains.

altitude: Height above sea level.

amphibian: An animal, such as a frog or salamander, that lives partly in water and partly on land.

archaeologist: A scientist who studies buried remains to investigate how people lived in the past.

atmosphere: The layer of air around Earth.

biome: A major division of the living world, distinguished by its climate and wildlife.

camouflage: A natural disguise that makes animals or plants look like their surroundings.

canopy: The rooflike layer of treetops in a forest.

carbon dioxide: One of the gases in air. Animals and plants produce carbon dioxide constantly.

charcoal: A solid black fuel made by partially burning wood.

climate: The pattern of weather that happens in one place during an average year.

cloud forest: A lush, misty forest found on mountains in the tropics.

cold-blooded: Having a body temperature that depends on the surroundings.

conifer: A type of plant that produces seeds in a cone and typically has needle-shaped leaves.

desert: A place that gets less than 10 inches (250 mm) of rain a year.

deforestation: The clearing of forest, usually carried out by cutting down or burning trees.

erosion: The gradual wearing away of land by the action of wind, rain, rivers, ice, or the sea.

epiphyte: A plant that grows on another plant and gets its water from the air or from rain.

equator: An imaginary line around Earth, midway between the North and South poles.

evaporate: To turn into gas. When water evaporates, it becomes an invisible part of the air.

evolve: To change gradually over many generations.

fertile: Capable of sustaining plant growth. Farmers often try to make soil more fertile when growing crops.

geyser: A jet of hot water or steam produced by volcanic activity.

hemisphere: One half of Earth. The northern hemisphere is the half to the north of the equator.

hibernation: A time of inactivity that some animals go through during winter. In true hibernation, the heart rate and breathing slow dramatically and the body cools.

ice age: A period in history when Earth's climate was cooler and the polar ice caps expanded. The last ice age ended 10,000 years ago.

invertebrate: An animal with no backbone, such as a worm.

irrigation: The use of artificially channeled water to grow crops.

limestone: A type of rock formed over thousands of years from the shells of tiny sea creatures building up on the seabed. Chalk is a type of limestone.

mammal: A warm-blooded animal that feeds its young on milk.

migration: A journey made by an animal to find a new home.

monsoon: A very rainy season in South Asia; or the wind that causes the rainy season.

nutrient: Any chemical that nourishes plants or animals, helping them grow.

oxygen: A gas in the air. Animals and plants need to take in oxygen so the cells of their body can release energy from food.

parasite: An organism that lives inside or on another organism and harms it.

plateau: An area of relatively flat land higher than its surroundings.

pollen: Dustlike particles produced by the male parts of a flower.

pollination: The transfer of pollen from the male part of a flower to the female part of the same flower or another flower.

predator: An animal that catches and eats other animals.

protein: One of the major food groups. It is used for building and repairing plant and animal bodies.

rain forest: A lush forest that receives frequent heavy rainfall.

rain shadow: An area where rainfall is low because a nearby mountain range obstructs rain-bearing winds.

reptile: A cold-blooded animal such as a snake, lizard, crocodile, or turtle that usually has scaly skin and moves either on its belly or on short legs.

sediment: Particles of mud, sand, or gravel carried by a river.

shrubland: A biome that mainly contains shrubs, such as the chaparral of California.

snow line: the level above which snow never completely melts on a mountain.

species: A particular type of organism. Cheetahs are a species but birds are not, because there are lots of different bird species.

temperate: Having a moderate climate. Earth's temperate zone lies between the tropics and the polar regions.

terrace: Part of a hillside that has been artificially leveled, usually for growing crops.

timberline: The line above which no trees grow on a mountain.

tropic of Cancer: An imaginary line around Earth 1,600 miles (2,600 km) north of the equator.

tropic of Capricorn: An imaginary line around Earth 1,600 miles (2,600 km) south of the equator.

tropical: Between the tropics of Cancer and Capricorn. Tropical places are warm all year.

tropical forest: Forest in Earth's tropical zone, such as tropical rain forest or monsoon forest.

tropical grassland: A tropical biome in which grass is the main form of plant life.

tundra: A biome of the far north, made up of treeless plains covered with small plants.

understory: A layer of plants between the ground and the canopy of a forest.

warm-blooded: Having a constantly warm body temperature. Mammals and birds are warm-blooded.

vapor: A gas formed when a liquid evaporates.

Further Research

Books

Alden, P. *National Audubon Society Field Guide to the Rocky Mountain States.* New York: Knopf, 1999.
Attenborough, David. *The Private Life of Plants.* Princeton, NJ: Princeton University Press, 1995.
Gerrard, John. *Mountain Environments.* Cambridge, MA: MIT Press, 1990.
Holmes, Don W. *Highpoints of the United States.* Salt Lake City: University of Utah Press, 2000.
Whiteman, C. David. *Mountain Meteorology.* New York: Oxford University Press, 2000.

Websites

Wild World: www.nationalgeographic.com/wildworld/terrestrial.html
(A National Geographic clickable world map, providing masses of information about the world's ecosystems.)
Peakware World Mountain Encyclopedia: http://www.peakware.com/encyclopedia/
(Includes descriptions of 1,700 of the world's highest peaks.)
World Mountains: http://mountains.egyo.com/
(Includes mountain mythology and true stories of mountain conquest.)

Index

Page numbers in *italics* refer to picture captions.

alpine tundra 26–28
Alps *7*, 32–33, 43
  plants 27, 28–29
Altiplano 13
altitude sickness 52
Andes 6, 18–19, 23
  plants 27, 31
antelope 43
Atacama Desert 13
avalanches 14, 16
Aztecs 47–48

bats 38
bears 36–37
birds *18*, 24, 34, 36, 37,
  39–40, 41, *41*
  hummingbirds 28, 31,
    34, 37, 38
  owls 58
  nutcrackers 33
bristlecone pines 21, *29*,
  30–31
bromeliads 24

caves 38
Cherrapunji 16
climates 10–17
clouds *12*, 17, 23–24, 25
cloud stripping 17
condors *18*, 41
conifers 24, 25–26, *25*
Cordillera, Western 9
cushion plants *28*, 29

dams 51–52, *51*, 60
deserts, mountain 13

earthquakes 56–57, 60
East Africa *7*, 44–45
  plants 31
edelweiss plants 28–29
epiphytes 24–25, *27*

espeletias 31
Everest, Mount 6, 10–11,
  54, 55

farming 48–49, 50
flowers 27
forests 21
  cloud *12*, 17, 23–24, *27*
  deforestation 51, 58, 59
  see also trees

geysers 8
glaciers 32, 59
global warming 59
gorillas 42
Great Basin 13, 30
groundsels 31, *31*
ground squirrels 39
guanacos *34*

hibernation 34, 36–37
Himalayas 7, *7*, 38, 49,
  54–55, 60, 61
  plants 20, 23, 29
  rain 13, 16
  temperature 11
  travel in 47
hummingbirds 28, 31,
  34, *37*, 38
hurricanes 10

ibex 43
Incas 18, 48, *50*

Kauai, island of 13, 16
krummholz 29

lakes, volcanic 21
lammergeiers *41*
leaves 25–26
lobelias 31, *31*, 45

Machu Picchu 18
McKinley, Mount 6, 9
Mexico 47–48

migration 38–39, 41
minerals and mining
  49–51, 58–59, *59*
mud slides 59

nutcrackers 33

orchids 28, 30
owls, northern spotted
  58

pandas 36
parks and preserves 61
permafrost 27
prairie dogs 36, 39
puna 27

quetzals 39

Rafflesia plants 21
rain 12–13
Rainier, Mount 8, 10
rain shadows 13–14
Rocky Mountains 6, 8–9
  plants 20, 26
  timberline 22

St. Helens, Mount 56,
  57–58
salamanders 38
sequoias, giant 26
shrubland 22, 29
Sierra Nevada 13, 22
snow 11, 14–16, *14*, 17
snowbells 27

snow leopards 34, 35, *35*,
  40, 43
snow line 16, 22
South Pole 14
sports 52–53, 61
sunshine 11–12

temperature 10–12, 26
  inversion 12
tepuis 30
Tibetan Plateau 49, 55
  animals 35–36, 40, 49
timberline 22–23, 26
torpor 38
trees 22, 23
  conifers 24, 25–26, *25*
  felling 51, 58
  leaves 25–26
  see also forests

valleys, climate of 12, 14
volcanoes 56, *56*, 57
vultures 41

Waialeale, Mount 13, 16
White Mountains,
  California 21
winds 10, 11, 12–14

yaks 35–36, 40, *40*, 49,
  50
Yellowstone 8, 14

zones, mountain plant
  life 22, 26

Picture Credits